ENERGY SCIENCE, ENGINEERING AND TECHNOLOGY

SOLAR READY BUILDINGS AND HOMES

PLANNING AND PROMOTION PRACTICES

ENERGY SCIENCE, ENGINEERING AND TECHNOLOGY

Additional books in this series can be found on Nova's website under the Series tab.

Additional E-books in this series can be found on Nova's website under the E-book tab.

ENERGY POLICIES, POLITICS AND PRICES

Additional books in this series can be found on Nova's website under the Series tab.

Additional E-books in this series can be found on Nova's website under the E-book tab.

ENERGY SCIENCE, ENGINEERING AND TECHNOLOGY

SOLAR READY BUILDINGS AND HOMES

PLANNING AND PROMOTION PRACTICES

WALTER I. COOK
AND
NATE M. JAMES
EDITORS

Nova Science Publishers, Inc.
New York

Library of Congress Cataloging-in-Publication Data

ISBN: 978-1-62081-573-1

Published by Nova Science Publishers, Inc. †New York

CONTENTS

PREFACE

Solar ready design is important if photovoltaic (PV) or solar hot water (SHW) technologies are to be installed on a building at any time during the building's lifespan. Solar ready also allows owners to take advantage of a changing energy market. The economics for installing solar on a new building are not always compelling, but, in the future, that picture could change with rising electricity prices and/or falling solar technology costs. It is important to recognize that solar ready itself does not reduce energy use or replace conventional energy with green energy. Solar ready does not contribute toward carbon emissions reduction until actual solar capacity is installed. Instead, solar ready has the potential to act as a catalyst for future installations and market expansion by reducing balance of system costs at the time of installation. In order to see this benefit, solar ready must become a widespread practice. This book explores the mechanisms for encouraging solar ready design and construction.

Chapter 1 - According to the U.S. Census Bureau and U.S. Department of Housing and Urban Development's January 19, 2011, report on new residential construction, 7.8 million new, privately-owned housing units were completed in 2010.[1] These homes, if not compatible with solar technology, represent a large barrier to widespread solar deployment. Once a structure is built, structural and solar access issues can prevent a solar project from being cost effective, and, in some cases, can make it entirely infeasible. Currently, when developers and solar installers assess buildings as potential solar sites, they sometimes dismiss buildings that are incompatible with a solar installation. While solar could be pursued, the potential solar production is significantly reduced, making payback periods longer and financing less appealing. Homes and commercial buildings built "solar ready" offer a solution to overcoming this installation barrier to renewable energy.

Chapter 2 - This document identifies the important aspects of building design and construction to enable installation of solar photovoltaic and heating systems at some time after the building is constructed. This document addresses photovoltaic (PV), solar hot water (ST), and solar ventilation preheat (SVP) systems.

In: Solar Ready Buildings and Homes
Editors: W. I. Cook and N. M. James

ISBN 978-1-62081-573-1

Chapter 1

SOLAR READY: AN OVERVIEW OF IMPLEMENTATION PRACTICES*

Andrea Watson, Linda Giudice, Lars Lisell, Liz Doris and Sarah Busche

ACKNOWLEDGMENTS

Thanks to Keyes and Fox, LLP; Amanda Bybee, Namaste Solar; Chris Klinga, Lighthouse Solar; James Critchfield, EPA; Misty Conrad, Sarah Truitt, Jason Coughlin, Sheila Hayter, Jesse Dean, and Rachel Romero, NREL, for their contributions to this report.

INTRODUCTION

According to the U.S. Census Bureau and U.S. Department of Housing and Urban Development's January 19, 2011, report on new residential construction, 7.8 million new, privately-owned housing units were completed in 2010.[1] These homes, if not compatible with solar technology, represent a large barrier to widespread solar deployment. Once a structure is built, structural and solar access issues can prevent a solar project from being cost

* This is an edited, reformatted and augmented version of the Technical Report NREL/TP-7A40-51296, U.S. Department of Energy, dated January 2012.

effective, and, in some cases, can make it entirely infeasible. Currently, when developers and solar installers assess buildings as potential solar sites, they sometimes dismiss buildings that are incompatible with a solar installation. While solar could be pursued, the potential solar production is significantly reduced, making payback periods longer and financing less appealing. Homes and commercial buildings built "solar ready" offer a solution to overcoming this installation barrier to renewable energy.

A solar ready building is engineered and designed for solar installation, even if the solar installation does not happen at the time of construction. The solar ready design features, if considered early in the design process, are typically low or no cost. Attention to building orientation, available roof space, roof type, and other features is key to designing solar ready buildings. The *Solar Ready Buildings Planning Guide*, published by the National Renewable Energy Laboratory (NREL) in 2009, details the technical processes for making buildings solar ready.[2] In cases in which it is not cost effective or feasible to install a solar system at the time of construction, creating a solar ready structure improves the cost effectiveness of solar when pursued at a later date. This eliminates barriers to future solar applications and facilitates market growth.

Solar ready design is important if photovoltaic (PV) or solar hot water (SHW) technologies are to be installed on a building at any time during the building's lifespan. Solar ready also allows owners to take advantage of a changing energy market. The economics for installing solar on a new building are not always compelling, but, in the future, that picture could change with rising electricity prices and/or falling solar technology costs. In some locations, installing SHW is now cost effective for new construction projects. Building a home or commercial building that is not solar ready exposes owners to the risk of not being able to take full advantage of future economic scenarios for solar electricity and hot water. Instead, the building owners have no option but to continue to buy conventional power or pursue solar with a lower return on investment due to the structural or solar access barriers on their buildings. While installing solar at the time of construction is encouraged, when this option is cost prohibitive, solar ready could be a standard and common alternative.

It is important to recognize that solar ready itself does not reduce energy use or replace conventional energy with green energy. Solar ready does not contribute toward carbon emissions reduction until actual solar capacity is installed. Instead, solar ready has the potential to act as a catalyst for future installations and market expansion by reducing balance of system costs at the time of installation. In order to see this benefit, solar ready must become a widespread practice.

This report will explore three mechanisms for encouraging solar ready design and construction: solar ready legislation, certification programs for solar ready design and construction, and stakeholder education. These methods are not mutually exclusive, and all, if implemented well, could contribute to more solar ready construction.

Solar ready legislation is a way to mandate the uptake of solar ready practices. The Solar Ready Legislation and Policy section of this report describes the pros and cons of such an approach and discusses potential interactions with other policies. Appendix B includes sample solar ready legislation.

Certifying solar ready may occur through a solar ready certification program or through integrating solar ready into existing programs, such as the U.S. Department of Energy (DOE) and the U.S. Environmental Protection Agency (EPA) ENERGY STAR program[3] or the U.S. Green Buildings Council (USGBC) Leadership in Energy and Environmental Design (LEED)[4] program. Certifying a building or home as solar ready can work toward the long-term goal of reducing costs and carbon emissions over time. Certification helps the building owner advertise solar ready status and signifies the potential of the structure to take advantage of solar technologies later in the building's life. In addition, certification helps raise awareness among property owners who may otherwise not have considered solar at all. However, certifying solar ready offers a reward for a measure that does not save energy, produce clean energy, or reduce emissions.

Education of stakeholders can occur in conjunction with solar ready legislation or certification or as a standalone effort. Education of stakeholders will be an important step in widespread adoption of solar ready practices.

This report is intended for local, state, and federal decision makers as an overview and discussion of tools and methods for promoting widespread solar ready building practices.

SOLAR READY AND COSTS

When a building owner attempts to install solar on an existing building that is not solar ready, the owner may experience the following:

- Structural or solar access barriers that prevent installing solar due to insurmountable costs
- A less attractive financial picture due to reduced solar production
- An increase in cost to make the building compatible with solar.

All of these scenarios may slow or stop the rate of solar installations, and all can be avoided through solar ready building practices. This section explores the costs of solar ready, which can be used as background information for discussing the methodologies of promoting solar ready throughout this report.

Table 1. Cost Prohibitive Scenario

Structural or Solar Access Issue	Potential Cost	Solar Ready Fix	Comments
Roofs unable to accept added load of PV or SHW system, especially if ballast system is only option	A cost-intensive roof reinforcement may be the only option, making a solar project cost prohibitive.	Design roofs for anticipated loads associated with a future solar installation during the initial building construction.	This varies by location and may be more common where roofs are not designed for snow loads or where wind loading is especially high.
Significant* number of roof obstructions	Roof obstructions may be numerous enough that the added cost of building around them or moving them, along with the loss of solar production, makes the project infeasible.	Group roof equipment on the north side of the roof or locate vents and other equipment to utilize non-rooftop space.	This will more likely make a project infeasible in the commercial sector. A third-party financier—crucial to making many commercial solar projects move forward—may dismiss projects lacking sufficient continuous roof space.
Significant* shading	Many things, including architectural features, other buildings, and vegetation, can cause significant shading. When the shading is very extensive and unavoidable, the economic case for solar is difficult to justify.	During the design phase, ensure that architectural features and the building's position on the site minimize shading. Consider landscape architecture that does not shade the roof in the future.	Shading may be more common in highly vegetated areas or places with high-density buildings, such as large cities. The effect of shading on PV power production is greater than that to SHW thermal production.

* The actual value of "significant" will vary by region, sector, or technology.

Cost Prohibitive Scenario

In some cases, installing solar on a pre-existing building is cost prohibitive. Structural barriers or solar access issues can create a scenario where it is not feasible to install solar. Table 1 highlights these issues, as well as the cost scenarios that could make a project infeasible, the potential fix with solar ready design, and additional comments.

An entity that could take advantage of a production-based local incentive would be more affected by production losses than one that could take advantage of an installed capacity-based incentive, although some installed capacity-based incentives do consider potential production.

Reduced Production Scenario

In many cases, a building with structural or solar access barriers to solar can still be compatible with solar. However, the structural or solar access issues can increase up-front system costs or reduce system output. Structural issues such as roof load limitations require the need for structural reinforcement, in turn increasing up-front costs and total system size. Daily or semi-annual shading and roof obstructions that either shade or reduce the available roof space for solar can reduce the output of a solar system. Likewise, a non-ideal building orientation reduces system output. The example below highlights the difference in production when roof orientation is less than ideal.

EXAMPLE: EFFECT OF ORIENTATION ON PV

A 10-kilowatt (kW) PV array in Golden, Colorado, facing south and tilted 25° can be expected to produce 14,304 kW hours (AC) per year with an annual energy value of $1,201.00. (The state average electricity price is 8.4¢ per kW hour.) The same system facing east will produce 11,997 kW hours (AC) per year with an annual energy value of $1007.75. A west-facing system will produce 10,999 kW hours (AC) per year with an annual energy value of $923.92. This represents a 16% to 23% reduction in PV production and cost savings when oriented 90° away from south. Table 2 presents a simplified financial analysis of this example.

Table 2. Effect of Orientation on Finances for a 10-kW PV Array in Colorado

Orientation	Installed cost	Annual solar energy production value (8.4¢)	Cost after utility rebate program of $2/watt (this rebate based on installed capacity, not production)	Approximate cost after incentive mix, about 45% of cost (e.g., income tax credit, accelerated depreciation)	Simple payback period
South	$60,000	$1,201.00	$40,000	$22,000	18.3 years
East	$60,000	$1,007.75	$40,000	$22,000	21.8 years
West	$60,000	$923.92	$40,000	$22,000	23.8 years

The example above illustrates that orienting a building to maximize solar production can shave years off the payback period for a solar installation. In the example, the utility rebate program is based on installed capacity, not system production. Some rebate programs degrade incentive levels for systems that are not optimally configured, but this degradation was not considered here. An incentive program based on production would further increase the simple payback periods for installations facing away from south. Similar calculations can be made to account for reduced array size due to roof obstacles or reduced production due to shading; however, these scenarios will vary greatly from site to site.

When confronted with a scenario in which expected solar production is reduced, some building owners may choose to pursue solar despite the less attractive economics, while others may choose not to proceed at all. Designing a solar ready building can prevent the production losses that negatively affect solar installation economics later in a building's life. Solar ready buildings designed with continuous roof space uninterrupted by roof equipment, minimal roof shading throughout the year, and a roof oriented on an east-west horizontal axis can increase production and shorten simple payback periods.

EXAMPLE: PV SYSTEM

In this example, the costs are calculated for making a residential building solar ready for a 10-kW PV system. For this example, there are three main measures associated with making the building solar ready that can increase the cost of construction. They include upsizing the electrical panel, installing conduit and wire from the roof to the electrical panel, and installing a combiner box on the roof in which to combine the wiring for the panels. The costs for these measures were estimated using the 2011 edition of *RS Means Construction Databook*. By these calculations, the building can be made PV ready for $1,729 at the time of construction. See Table 3 for the cost breakdown of the measures.

If there were no solar ready preparations made at the time of construction, the same preparations would cost $4,373 at the time of solar installation. The measures would be more costly after construction for a number of reasons. Installing the measures during construction could be completed by a general contractor already on the site instead of requiring a team of solar contractors to travel to the site. This results in a 75% savings in the labor costs due to reduced travel time. In this example, vent relocation and inhibited roof exposure require the solar panels to be installed on multiple pitches. Both of these additions result in additional expenditures. The cost of the materials was assumed to be the same whether the work is done during or after construction. In this example, the measures completed during construction saved $2,644.

Table 3. Cost to Make a Building Solar Ready[5]

	During Construction			After Construction		
Measures	**Equipment**	**Labor**	**Total**	**Equipment**	**Labor**	**Total**
Increase size of electrical panel	$459	$480	$ 939	$459	$1,200	$1,659
Run conduit	$374	$416	$ 790	$374	$1,040	$1,414
Relocate vents	n/a	n/a	n/a	-	$300	$300
Install panels on multiple pitches	n/a	n/a	n/a	-	$1,000	$1,000
Total	**$833**	**$896**	**$1,729**	**$833**	**$3,540**	**$4,373**

Incremental Cost of Solar Retrofit

There is always a cost associated with retrofitting a building to accommodate solar. With the proper solar ready preparations, these measures may cost less if done at the time of building construction. The examples below consider cost differences between preparing a building for solar during initial construction and after initial construction. Quantifying these incremental costs is difficult because of site variations. The scenarios below are examples of cost reduction scenarios, but these examples cannot be extended to every situation.

In this example, implementing the solar ready measures during building construction costs 60% less than the measures would cost after construction.

Example: Solar Hot Water System

In this example, the costs are calculated for making a residential building solar ready for a three-panel, 96-ft^2, closed-loop solar thermal system. It is assumed that the system utilizes .75-inch tubing, and that the existing hot water system requires the addition of "stub-outs" in order to allow the SHW system to be incorporated without depressurizing the domestic hot water system. For this system, there are three main measures that will increase the cost of the building construction. They include installing copper pipes from the roof to the utility room, adding tee joints to the existing system where the SHW system can be plumbed in, and adding SHW mounting hardware on the roof of the building. By implementing these measures during construction, the building can be made SHW ready for $1,588.

Table 4 shows the cost breakdown of the measures. If there were no solar ready preparations made at the time of construction, the same preparations would cost $4,645 as a retrofit. The measures would be more costly after construction for a number of reasons. Installing the measures during construction could be completed by a general contractor already on the site instead of requiring a team of solar contractors to travel to the site. In this example, the result is a 72% savings in the labor costs due to reduced travel time. In this scenario, there are vents that need to be relocated and roof exposure that cannot be optimized, resulting in panels that need to be installed on multiple pitches. Both of these additions require additional expenditure.[6]

The cost of the materials was assumed to be the same whether the work is done during or after construction. In this example the measures completed during construction saved $3,057.

Table 4. Cost to Make a Building SHW Ready[7]

	During Construction			After Construction		
Measures	**Equipment**	**Labor**	**Total**	**Equipment**	**Labor**	**Total**
Add mounting hardware SHW	$8.00	$20.00	$28.00	$8.00	$50.00	$58.00
Pipes to roof	$407.00	$1,109.00	$1,516.00	$407.00	$2,773.00	$3,180.00
Stub-out pipes	$2.00	$43.00	$45.00	$2.00	$106.00	$108.00
Relocate Vents	n/a	n/a	n/a	–	$300.00	$300.00
Install panels on multiple pitches	n/a	n/a	n/a	–	$1,000.00	$1,000.00
Total	**$417.00**	**$1,172.00**	**$1,589.00**	**$417.00**	**$4,229.00**	**$4,646.00**

In this example, implementing the solar ready measures during building construction costs 66% less than the measures would cost after construction.

SOLAR READY LEGISLATION AND POLICY

This section discusses solar ready legislation or policy tools that may be useful in promoting solar ready. In preparing the sample legislation in Appendix B, NREL worked with Keyes and Fox, LLP, to evaluate a variety of different approaches that have been taken by states and municipalities in requiring new building construction to be solar ready. Keyes and Fox, LLP, drafted the sample legislation in Appendix B to provide a starting point for jurisdictions interested in creating solar ready legislation.

There are a number of policies that could be considered when designing solar ready legislation and several that can act as standalone policies to promote solar ready. The following solar ready legislation discussion explains some of the choices reflected in the sample presented in Appendix B. This section also highlights some complementary policies and options that could strengthen the effectiveness of solar ready legislation.

Requiring solar ready construction through legislative or regulatory mechanisms has certain advantages, but may work best when combined with some of the other tools discussed below, such as certification programs or education strategies. One of the biggest challenges of legislation is the ability to craft stringent rules requiring buildings to be solar ready while also allowing developers to achieve established requirements in a manner well suited to a particular building type and a developer's design goals. Working in coordination with developers and builders helps make legislation more politically feasible. It is also important that legislation enables the use of the most cost- and resource-efficient technologies as they evolve.

Currently, several jurisdictions have put in place some sort of policy that promotes solar ready construction (see Appendix C). However, there is currently little data on the effectiveness of these policies by metrics. More data, such as the number of new solar ready structures as a result of policy or the increase in the number of solar installations, needs to be collected on existing policies. Future policies should also plan for data collection and verification of policy effectiveness.

Mandating Solar Ready Buildings or a Solar Ready Option

It is important to highlight the difference between solar ready legislation that *mandates* solar ready versus legislation that requires a solar ready *option*. A number of states have required that the option of solar readiness be expressly given to homebuyers at the time of construction.[8,9,10] Instead of imposing a requirement on new construction to be solar ready, this option focuses on educating the homeowner about solar technologies, providing them with the information to decide if the added cost of building solar ready is of value to them when they consider how future energy costs will impact their budget. Conversely, the sample legislation in Appendix B has been drafted to require that new residential and commercial construction be solar ready. One of the limitations of the "option" approach is that relatively few homes and commercial buildings are constructed with the ultimate buyer in mind. Unless the developer or builder will be the ultimate owner of a structure, choosing the solar ready option may not be appealing. Consideration of the variety of owners of a building over its usable lifetime increases the appeal of solar ready construction so that a long-term tenant or future owner could easily install solar.

Installing a solar system on a solar ready building can be more cost effective than installing solar on a building that is not designed to accommodate solar. Some features, such as orientation or shading, may not be reasonable to modify at a later date (see "Solar Ready and Costs" section). In addition, requiring a building to be constructed solar ready from the start may provide additional motivation for initial building owners to look more seriously at solar options in connection with the initial investment in constructing a residential or commercial building.

COLORADO HB 1149

In Colorado, HB 1149 was enacted in May 2009, requiring builders of single-family homes to offer solar as a standard feature to all prospective homebuyers. Builders are required to give the buyer the option to either have a PV system or a SHW system installed on the new home, or to have all the necessary wiring and/or plumbing installed so that the homeowner can easily add a solar system at a later date. The builder must also provide the buyer with a list, maintained by the Governor's Energy Office, of every solar installer in the area so the buyer can obtain expert help in determining if the home's location is suitable for solar and what the estimated cost savings would be (General Assembly of the State of Colorado: Section 1, Article 35.7 of title 38, Colorado Revised Statutes).

Along with mandating solar readiness for new construction, it may be effective to also require that future building owners be informed of options for having a solar system installed at the time a building is first constructed. Colorado and New Jersey have instituted requirements of this sort.[11,12] Software tools, such as In My Backyard,[13] can provide information to an owner/developer on what a building's exposed surface area could produce with a PV array installed. Likewise, the NREL SolOpt[14] tool can be used to determine if solar hot water or PV makes more sense for a given amount of roof space on a particular building. Developers could be required to provide such information to educate buyers on the potential benefits of installing a solar system.

Solar Ready Sample Legislation

The sample legislation in Appendix B has been drafted such that state or local agencies have the opportunity to consider local conditions and policy appetite in promulgating detailed regulations for solar ready building requirements. Although detailed requirements could simply be included in statutory language itself, there are a number of advantages to having the detailed specifications included in regulations. Principally, regulations are generally easier to update and modify over time. Given that technologies change, the ability to update technical requirements without action from the legislature is important to providing necessary flexibility. However, legislators can include more prescriptive language about technical requirements in the legislation if preferred. The technical requirements that NREL currently recommends have been provided in footnotes to the sample legislation for further guidance to implementing agencies.

If time and resources are not available at an implementing agency level, an effective solar ready statute can be drafted that avoids the need for further regulatory action by including specific requirements directly in a statute itself.

Commercial and Residential Considerations

The sample legislation in Appendix B applies to both commercial and residential construction. Although many of the existing solar ready mandates adopted in the United States only apply to single-family residential buildings, there are similar cost challenges associated with modifying commercial construction that make extension of these mandates to such buildings a sensible approach. In particular, commercial buildings often have flat roofs that require ballast-mounted systems. Commercial building roofs need to be designed to withstand the added weight of solar systems. The varied nature of commercial buildings requires more detailed regulations to be adopted to accommodate a range of building types. States may also want to consider whether multi-family residential and multi-unit commercial buildings warrant separate regulatory guidance.

Sample Legislation and Incentives

The sample legislation in Appendix B includes a provision to make it clear that compliance with solar ready mandates is not required for homes constructed prior to the statute to access any available incentives for solar installations. A new building that complies with the statute should also be able to access incentives under the theory that the costs of installation will be less.

Green Building Codes

Solar ready requirements can be a low-cost and effective addition to existing green building codes. Jurisdictions with green building codes using a points-based system may consider assigning points for aspects of solar ready construction, while providing additional points for homes that have actual solar systems installed. The text box highlights the residential rating system used in Pima County, Arizona. While this program is designed to promote solar ready and ultimately increase the number of solar installations, preliminary data needs to be collected on whether this program is effective or if most developers choose not to comply.

Pima County Regional Residential Rating System – Tucson, Arizona[15]

The Pima County Regional Residential Rating System was developed to promote the construction of sustainable homes specifically designed for Pima County. Homes are rated and awarded points for environmentally responsible criteria including location, lot design, and development as well as resource and energy efficiency. Solar ready (SHW and PV stub-outs) qualifies for points within this system. There are four levels of certification available that range from Bronze (75–100 points) to Emerald (over 160 points). This voluntary, no-cost program is combined with the permit and inspection process. The procedure for obtaining the rating includes looking at site development and resource, energy, and water efficiency. Once one completes the procedure of complying with the requirements of the program, the building receives a certificate indicating the award level and becomes eligible to receive a building plaque to commemorate the achievement. For more information, see www.pimaxpress.com/green.

Homeowners' Associations

Jurisdictions may consider including language that addresses how solar ready requirements interact with limitations imposed by homeowners' associations or common interest communities (CICs) rules. Although homeowners' associations cannot trump requirements imposed by state law, homeowners' associations that adopt rules directly or indirectly prohibiting the installation of solar equipment may undermine the intent of requirements that buildings be constructed solar ready. For example, states and localities include provisions that explicitly prohibit homeowners' associations from implementing restrictions that prevent homeowners from installing solar energy equipment on their own property (e.g., Arizona[16]). In this case, jurisdictions may consider expressly preempting homeowners' associations from prohibiting solar installations when enacting solar ready legislation. In Colorado, HB 08-1270 revised an existing law so that installation of renewable energy or energy efficiency equipment cannot be prohibited.[17] The bill prevents CICs from making rules that would explicitly prevent the installation of such equipment; however, the bill allows reasonable restrictions based on aesthetic considerations, safety requirements, and—in the case of wind-electric generators—noise.[18]

Session Laws of Colorado 2008, Second Regular Session, 66th General Assembly

Concerning the removal of restrictions on the use of energy efficiency* measures in connection with real property.

38-30-168. Unreasonable restrictions on renewable energy generation devices - definitions. (1)(a) A covenant, restriction, or condition contained in any deed, contract, security instrument, or other instrument affecting the transfer or sale of, or any interest in, real property that effectively prohibits or restricts the installation or use of a renewable energy generation device is void and unenforceable.

Historic Structures

The sample legislation in Appendix B only applies to new construction, so it is not likely to impact existing historic structures. However, new construction within a designated historic district may be under restrictions that inhibit the ability of a developer, or eventual owner, to install a solar system. As with homeowners' associations, it may be wise to consider expressly preempting any such restrictions or to provide for an exemption for particular circumstances.

Solar Access Laws

Another potential companion provision may be the creation or enhancement of solar access rights. There is an exception from compliance in the sample in Appendix B for buildings that do not have sufficient solar resources to warrant the addition of solar readiness at the time of construction; however, this exception does not protect a building that may have sufficient solar resources at the time of construction from future obstructions that may diminish access to solar resources. To ensure that initial investment in solar ready construction maintains its value over time, it may be worth considering companion provisions to enact solar access protections.

Nearly 80% of the states and a number of localities have implemented some type of a solar access law. As these policies are designed to meet the unique needs of the constituents, they vary greatly. For example, some of these policies allow parties to voluntarily enter into an agreement to protect the solar easement rights of either or both parties (e.g., Georgia[19]). Other policies require that solar access is not impeded (e.g., the City of Sebastopol, California[20], can place restrictions on vegetation and building construction if it impedes on solar easements).

Like many solar ready policies, data is needed on the effectiveness of these laws to protect solar access and encourage solar installation.

Roof Warranties

The requirements outlined in Appendix B require that roofs be designed to accommodate installations of solar PV systems and/or SWH systems. In light of this, it is important that roof warranties accommodate the ultimate installation of a solar system without the fear of a warranty being voided by the installation of a system. Jurisdictions adopting solar ready mandates may want to consider including requirements that will ensure that roof warranties are compatible with, and not voided by, the proper installation of a solar system.

Sample of Solar Ready Requirements or Practices by Jurisdiction

Appendix C summarizes the jurisdictions that have adopted solar ready into existing legislation, codes, or ordinances. The DSIRE website was a primary source for these excerpted entries.[21] A few others were obtained through direct e-mail or telephone contact with various jurisdictions. Of the examples cited, most require measures related to PV and SHW systems. Technical requirements range from SHW stub-outs to full PV site/situational plans (see Tucson, Arizona). It should be noted that there appear to be a number of jurisdictions where elements of solar ready are practiced, but not mandated (see the City of Boulder or Aspen/Pitkin County).

Data on the effectiveness of these different approaches to solar ready would be useful for other jurisdictions looking to adopt similar policies. While outside of the scope of this report, compiling information on participation, increased number of solar ready structures, and increased number of solar installations as a result of these policies would help jurisdictions choose the most appropriate policy for their area. Similarly, new policies put in place should consider how to evaluate the effectiveness of those policies in promoting solar ready and solar installations.

Incentives

Incentives are a mechanism used by governments for the purpose of encouraging the use of technologies considered to have public benefits. These can be financial incentives that reduce the cost of clean energy technologies, which are typically advanced, higher-cost technologies. Examples of financial incentives include tax credits or deductions, rebates, and reduced interest rates on loans. These increase access to clean energy technologies into broader markets. Alternatively, incentives can be non-financial, in which a benefit other than money is used to encourage the use of technologies. Examples of non-financial incentives are expedited permitting and “density bonuses,” which allow developers to increase the maximum allowable development on a property in exchange for acting in the public good.[22] While both types of incentives could be used to promote solar ready building, neither is currently used to do so. A literature review found no current incentives for solar ready construction.

Making incentives available could raise the profile of building solar ready as well as encourage developers to do so. Such incentives would need to be carefully structured. Creating financial incentives for solar ready buildings is complicated because the buildings result in a potential for clean energy development, not actual clean energy production. Therefore, calculating the expected public good resulting from solar ready development is challenging. Non-financial incentives may be more straightforward, as they result in little direct cost to the government, so justification for the implementation of the incentive may be more politically palatable.

Conclusion

Legislating solar ready construction is one method for promoting solar energy deployment. As opposed to the other methods discussed next, this method effectively requires developers to consider solar ready design. Pursuing solar ready through legislation should take into account a diverse group of stakeholders. Combined with training and educational tools, solar ready legislation could effectively increase solar ready design practices. Other methods for implementing solar ready may prove faster to initiate and offer more flexibility to stakeholders.

SOLAR READY CERTIFICATION

Certification Programs

This section discusses the use of certification programs as an option for advancing solar ready implementation. Solar ready certification could act as a standalone program or be integrated into existing programs such as ENERGY STAR and LEED. This section will discuss both approaches.

Solar Ready Certification's Potential Benefits

In general, certification is used in green building to promote a standard and ensure desirable characteristics such as design, quality, environmental friendliness, safety, reliability, efficiency, and interchangeability.

If it is not cost-effective or feasible to install solar at the time of construction or retrofit, creating a solar ready structure will save money and time when solar is pursued at a later date. This would eliminate barriers to future solar applications and facilitate market growth. In addition, certifying a building or home as solar ready helps the building owner advertise solar ready status and signifies the potential of the structure to take advantage of solar technologies later in the building's life. Lastly, it helps raise awareness among property owners who may otherwise not have considered solar from the outset.

To summarize the potential benefits, certification of solar ready:

- Ensures a certain level of quality. Currently, any property could be advertised as solar ready, but the lack of a standard definition or technical requirements prevents this claim from carrying any weight.
- Signifies the opportunity to take advantage of changing electricity and solar markets.
- Allows building owners an easy avenue for advertising the status of their properties or homes.
- Provides a measurable metric.
- Is voluntary and may be more palatable to building owners or developers than legislation.
- Can serve as a form of communication, awareness, and education on solar.
- May promote solar ready as a catalyst for market expansion by overcoming barriers that prevent solar projects from going forward.

Challenges of Solar Ready Certification

One of the clear disadvantages of a solar ready certification program is that it does not guarantee installation of solar systems. If this becomes the norm, then the perceived value of solar ready programs could be diminished. Furthermore, solar ready itself does not reduce energy use or replace conventional energy with green energy, which raises the question as to whether solar ready should even be rewarded with a certification.

When certification schemes are adopted, they regularly cite statistics regarding improved standards, benefits, and financial savings. Yet, sometimes they can fail to reach their potential. For certification schemes to work, they require proper design and management to achieve the intended results.[23]

A Solar Ready Certification Label

Similar to the ENERGY STAR or LEED label, guidelines could be established for achieving a solar ready label that stands separate from these existing programs.[24] An easy way to make sure a building is solar ready would be to find out if it has been solar ready certified. This label could be government based like ENERGY STAR or nonprofit backed like LEED. Local jurisdictions could also consider creating a solar ready certification and label. Any building may be able to earn a solar ready certification label if the building meets the established requirements.

There are several advantages of a standalone solar ready label that is specific to the achievement of being solar ready. If solar ready is incorporated into a more comprehensive green building program, making a structure solar ready would likely be one of several choices of green building measures one could choose from to obtain certification. This could inherently dilute the number of structures that become solar ready. A solar ready label would only be applied to buildings that were indeed solar ready. Furthermore, a standalone solar ready label would be positioned to gain brand recognition, raising awareness about solar installations and solar ready buildings. This brand could be developed into a sticker or plaque that could be put on infrastructure such as electrical panels that have room for solar interconnection, conduit, or pipes to signify solar ready status to future users.

A disadvantage of having solar ready as a standalone label as opposed to part of an existing certification program might be the more limited visibility and participation. However, it remains a good option for local jurisdictions or other entities wishing to emphasize the importance of solar ready. Additionally, too many certifications and label programs for this sector could complicate the market to the point where the solar ready label has little value.

Solar Ready Certification Status

ENERGY STAR

From a review of existing solar ready initiatives, there does not appear to be a certification program dedicated exclusively to solar ready practices in the United States. However, at this writing, actions have been taken to incorporate solar ready into ENERGY STAR and LEED certification programs.

ENERGY STAR is a DOE and EPA program focusing on improving energy performance in buildings as a method of reducing greenhouse gas emissions. ENERGY STAR recognition is available as an indicator of superior energy performance. Projects can alternatively achieve Designed to Earn ENERGY STAR to recognize architects' intent that buildings operate at superior energy efficiency when built. Once the buildings are occupied and operating, they can earn the ENERGY STAR label.[25]

The EPA has developed renewable energy ready home specifications to incorporate into the ENERGY STAR Concept Home Version 4 program requirements (ENERGY STAR Version 3 will be fully adopted as the mandatory base program guidelines by January 1, 2012, followed by Version 4).[26] The content of Version 4 was rolled out before 2012, but it will be an optional guideline until it is completely phased into the program, replacing Version 3 (date of complete phase-in to be determined).[27] Currently, PV and solar thermal are the only two technologies addressed by the renewable ready home specifications.

In early 2012, the EPA plans to pre-release the renewable energy ready home specifications as an educational resource and to provide the opportunity for the ENERGY STAR builder community to give feedback. During the pre-release period, homes meeting the renewable energy ready specification would not be eligible as renewable energy ready homes for recognition under any ENERGY STAR program guidelines.

By incorporating renewable energy ready homes into ENERGY STAR, the EPA is maintaining a focus on energy efficiency while rewarding homes that are prepared to take the next step to solar energy. EPA designers of the solar ready guidelines see a large opportunity to educate both the home building community and homeowners through the solar ready addition. Homebuilders can learn simple design and building practices that make a home solar ready. Upon receiving information about their solar ready home, homeowners can then learn about the opportunity to install solar systems.

In the ENERGY STAR renewable energy ready home checklist, applicants will have to meet a number of requirements in the following categorical areas:

- Building Array/Site Assessment
- Structural and Safety Considerations: Photovoltaic (PV) or Solar Hot Water (SHW)
- Renewable Energy Ready Home Infrastructure
- Homeowner Education.

Leed

The Leadership in Energy and Environmental Design (LEED) Green Building Rating System is a green building certification offered by the nonprofit U.S. Green Building Council (USGBC).[28] LEED is a building certification process that looks at various aspects of green building and gives recognition to structures that meet certain standards. LEED has recently developed a draft titled "EA Credit: Solar Ready Design" as part of its Homes Rating System with the designated intent to "reduce energy consumption and greenhouse gas emissions by designing to maximize opportunities for solar design." [29] The draft emphasizes the following areas for solar ready design:

- Infrastructure requirements (hardware, raceway for PV, plumbing for SHW)
- Design approval by a PV or SHW contractor
- Minimum roof space and orientation
- Informational material for occupant.

At this writing these draft changes to LEED Homes Rating System are in the public hearing phase and may be further modified before formally accepted into practice. A link to the document that was open for comment can be found at the USGBC website.[30]

Conclusion

Both ENERGY STAR and LEED programs promote green building practices by offering a reward for incorporating specified characteristics into a building. The reward, or certification, offers a value to the owner for a variety of reasons. This may be from an increased market value of the property or the ability of the owner to meet federal or local mandates, or because the certification may serve as a symbol of environmental responsibility.

As opposed to mandating solar ready buildings through legislation, certification programs or guidelines encourage building owners to adopt solar ready practices by offering a reward or recognition. A drawback of incorporating solar ready into certification schemes is that it rewards building owners for building attributes that don't reduce electricity consumption or generate green electricity. Nevertheless, a building stock of solar ready buildings would reduce installation barriers and likely increase solar production. Certification of solar ready buildings is a reasonable way to encourage solar ready design.

STAKEHOLDER EDUCATION AND COMMUNICATION METHODS

Stakeholder education represents a critical component of expanding the implementation of solar ready building practices. Whether in conjunction with solar ready legislation or certification programs, or as a standalone entity, education around solar ready is important to promoting widespread solar ready buildings. Solar ready buildings themselves further serve as conduits for solar education. This section explores multiple approaches to providing solar ready education.

Audience

Owners and real estate developers are the stakeholder groups most likely to benefit from a solar ready home or building. An owner with a solar ready home is positioned to take advantage of falling prices of solar technologies. At some point in the future, an owner can install a solar array to hedge against rising electricity prices. Likewise, real estate developers can advertise solar ready homes as a green investment in the future. Owners and developers ultimately decide how a building is designed and built. Education programs should emphasize the merits of equipping a house for solar at the time of construction as a best practice. However, in situations when solar installation at the time of construction is not feasible or is cost prohibitive, solar ready should be presented as the next best option.

Engineers and architects must understand the simple principles of solar ready design. Often, these principles can be applied without changing the cost or quality of the building design and should be considered good building practices. Solar ready is an added-value feature that engineers and architects can offer owners in their design that requires little up-front cost. Furthermore, knowledge of solar ready principles prepares architects and engineers to design buildings that meet mandates or certification requirements.

Methods of Delivering Education

Considering the large number of homes built in the United States each year, solar ready education of key stakeholders could have a significant impact on the future of solar in this country. Education and communication strategies must reach a large number of people in diverse fields. The discussion below explores several methods for reaching target audiences. Local governments, federal agencies, or nonprofit organizations could implement these methods.

Education Combined with Legislation or Policy

A strong approach for promoting solar ready is the combination of solar ready policy with an educational campaign. In this approach, the mandate requires developers, architects, and engineers to consider solar ready, while the educational campaign ensures they understand the best practices. Pairing education with a mandate motivates stakeholders to participate in education or training programs or to use available educational tools. Designing an education program in parallel to developing mandates may also help make mandates more palatable to the groups most affected.

Education Combined with Certification

A certification program promotes solar ready building practices by offering a label or certificate and accompanying educational aspects. These aspects may include training programs, robust websites, and public awareness campaigns.

SOLAR READY TRAINING – BOSTON, MASSACHUSETTS

The City of Boston's Department of Neighborhood Development (DND)[31] has developed design standards for new construction and the rehabilitation of existing buildings. These ensure that all projects conform to current applicable regulations and promote cost effective, environmentally responsible quality design. DND requires that applicants for subsidized funding for building development be solar ready. DND has worked with the Solar Boston Initiative to incorporate solar into the housing portfolio through the Green Affordable Housing Program (GAHP). In conjunction with the solar ready requirement, the Green Affordable Housing Program offers specialized training sessions on Integrated Design and Renewable Energy and Energy Efficiency to help developers meet the solar ready and other green building requirements.

The EPA and USGBC have both had success promoting their respective certification programs, ENERGY STAR and LEED. USGBC has a robust *education program* that parallels their certification program. Education is designed for those who are implementing LEED projects or seeking LEED certification

USGBC provides a curriculum that is taught through a wide network of education providers. Through this infrastructure, USGBC is able to offer hundreds of courses worldwide. Achieving LEED certification requires an educated workforce. In turn, USGBC provides training to the workforce that completes LEED projects. The projects themselves then act as an educational opportunity to the public that works or lives in high-performance green buildings. In the LEED Home Rating System draft, providing information about solar ready to the occupant is required.

SOLAR READY GUIDELINES – THE TWIN CITIES, MINNESOTA[33]

The Twin Cities, Minnesota, have released two location-specific documents for use by public agencies, community organizations, the nonprofit and for-profit development communities, and owners, architects, builders, and contractors. The *Solar Ready Building Design Guidelines* explain solar ready and outline guidance for designing and building a solar ready structure. The *Solar Ready Construction Specification* provides an implementation methodology for achieving solar ready buildings. The guidelines, released by the Twin Cities Solar America Cities Program in 2010, "… address specific site planning, building form, space planning, roofing, and mechanical and electrical issues to be considered in the design of solar ready buildings." The accompanying specification document "… establishes responsibilities and procedures for implementing these requirements during the construction phase of the building process." Both of these documents represent comprehensive education on solar ready building principles for a targeted audience in a targeted location. Supported by the city through the *Solar America Cities Initiative*, these documents will help overcome "major barriers to the installation of solar energy systems [which] is the traditional design of [the] building stock" in the Twin Cities.

Likewise, the EPA has recognized the educational potential of introducing the renewable energy ready home concept into ENERGY STAR. Ultimately, the EPA plans to incorporate the renewable energy ready home into the

ENERGY STAR Concept Home Version 4. Because the Concept Home Version 4 will likely not be available for several years, the EPA plans to pre-release the renewable energy ready homes standard. This allows it to be available as an educational tool and resource for ENERGY STAR builders while encouraging the community to provide feedback on the specifications.

While the EPA recognizes that it will be difficult to track how many renewable energy ready homes actually result in installed renewable energy projects, the program nonetheless serves as an opportunity to educate homeowners about renewable energy. Like ENERGY STAR or LEED, local or standalone solar ready certification programs or labels can be conduits for education and communication about solar ready. Incorporated in ENERGY STAR or LEED, solar ready will reach a wider audience. A potential drawback of this approach is the dilution of the concept by couching it with other green building measures. On the other hand, this emphasizes the importance of pursuing energy efficiency concepts in parallel with renewable energy.

Education Combined with Incentives

One approach may be to make educational training mandatory in order to receive an incentive or benefit. This benefit could be monetary, access to a streamlined process, or eligibility for a certification.

Stakeholder Education

Appendix A lists organizations that can act as conduits for educating the stakeholders on solar ready construction. These organizations often sponsor conferences, workshops, robust websites, and newsletters, and many have a database of stakeholders. Promoting solar ready concepts through these organizations may be an effective way to reach front-end decision makers who can pave the way for solar development.

Guidelines

Some jurisdictions have taken an approach to promoting solar ready that neither mandates compliance nor rewards participants with a label or certification. Solar ready guidelines fall into this third category. Solar ready guidelines offer specific information and even specifications that can be used to educate stakeholders about solar ready and aid stakeholders in implementation. Solar ready guidelines provide the information needed to construct solar ready buildings for those who are self-motivated to do so. When promoted by a visible entity or program, guidelines can encourage stakeholders to consider building solar ready when they might not have otherwise.[32] Likewise, the Twin Cities of Minneapolis and St. Paul, Minnesota, have pursued a solar ready building program based on location-specific guidelines and specifications.

Conclusion

Solar ready itself does not reduce energy use or create clean energy. Nevertheless, solar ready building practices are needed to reach the full potential of solar deployment. Without forethought on incorporating solar into design, buildings may be incompatible with solar due to roof structure or excessive shading. In these cases, retrofitting the roof or removing shading elements is cost prohibitive. Furthermore, higher up-front costs due to structural adaptations and production losses caused by less than optimal roof orientation, roof equipment, or shading will lengthen payback periods, making solar more expensive. With millions of new buildings constructed each year in the United States, solar ready can remove installation barriers and increase the potential for widespread solar adoption.

There are many approaches to promoting solar ready, including solar ready legislation, certification programs, and education of stakeholders. Federal, state, and local governments have the potential to implement programs that encourage solar ready and in turn reduce barriers to solar deployment. With the guidance in this document and the examples of jurisdictions and organizations already working to promote solar ready building practices, federal, state, and local governments can guide the market toward solar ready implementation.

APPENDIX A. TRADE ORGANIZATION RESOURCES

American Institute of Architects (AIA)

The AIA Board of Directors recognizes the need to prepare their members to respond to the challenge and opportunity facing the profession in environmentally responsible projects. The AIA offers a number of forums and programs by which industry professionals can learn about the latest developments. The AIA offers a continuing education program dedicated to sustainability in which solar ready modules may be incorporated.

American Solar Energy Society (ASES)

The American Solar Energy Society provides a conduit to integrate solar ready into solar energy education and communication. While the membership is often focused on solar installation, an additional emphasis on solar ready helps ensure that opportunities to make buildings solar ready will be cost effective into the future.

National Association for Industrial and Office Properties (NAIOP)

NAIOP, the Commercial Real Estate Development Association, is the leading organization for developers, owners, and related professionals in office, industrial, and mixed-use real estate. NAIOP provides industry networking and education, and advocates for effective legislation on behalf of its members. NAIOP advances responsible, sustainable development that creates jobs and benefits the communities in which members work and live. The NAIOP Center for Education is the principal learning resource for the commercial real estate professional. It is organized as a system of *continuing education*, which includes webinars, courses, certificate programs, and chapter education partnership.

National Association of Home Builders (NAHB)

The National Association of Home Builders (NAHB) is one of the largest trade associations in the United States, with more than 800 state and local associations. About one-third of NAHB's more than 175,000 members are home builders and/or remodelers. The home builder members of NAHB construct approximately 80% of new homes each year in the United States. The various NAHB groups analyze policy issues and economic and consumer trends as well as educate and communicate industry news to the public. The NAHB organizes the International Builders' Show, which is one of the largest conventions in North America for the residential and light commercial construction industry. NAHB publishes *Nation's Building News*, a free weekly e-newspaper distributed to nearly 200,000 recipients. The NAHB hosts educational seminars at the International Builders' Show, which offered 900 courses to 6,000 conference participants in 2008.

U.S. Green Building Council (USGBC)

The USGBC has diverse resources and multiple formats for LEED and green building education and knowledge. Solar ready could potentially be adapted within existing programs. The USGBC offers a number of educational forums, including:

- USGBC's LEED curriculum and workshops
- Reference guides
- Study guides
- Podcasts
- USGBC live chats
- Private LEED workshops.

Appendix B. Sample Solar Ready Legislation

1. Definitions – As used in the [*name of act or chapter*],
   A. "New Construction" includes any newly constructed residential or commercial building that requires a building permit to proceed.
   B. "Developer" means any person or company who constructs residential or commercial buildings.
   C. "Department" means [*name of state agency who will undertake tasks identified below*].
   D. "Solar Energy System" means any system that uses solar energy to provide all or a portion of the electrical needs of a residential or commercial building.
   E. "Solar Hot Water Heater" means any system that uses solar energy to heat water for use in a residential or commercial building.
   F. "Substitute Renewable Energy System" means any system that uses renewable energy resources, including wind and geothermal heat, to provide for all or a portion of the electrical needs of a residential or commercial building.
   G. "Substitute Renewable Hot Water System" means any system that uses renewable energy resources, including geothermal heat, to

provide for all or a portion of the hot water needs of a residential or commercial building.

2. New Construction Shall be Built Solar Ready
   All new construction shall be built to accommodate the installation of a solar energy system and solar hot water heater in accordance with regulations adopted by the Department and in accordance with the following sections.
3. Solar Energy System Readiness Standards
   A. The Department shall promulgate regulations by [date] that establish minimum standards that must be met for new construction to accommodate a solar energy system.
   B. In drafting the regulations, the Department shall take into account existing building code requirements and compliance costs. The Department shall also consult with scientists, engineers, professional societies, and government agencies with relevant expertise in solar energy systems and building construction.
   C. At a minimum, the Department shall include requirements for:
      i. Static load roof strength, with a requirement that roofing where solar equipment could be placed be capable of supporting a minimum of 6 lb/ft^2;
      ii. Placement of non-solar related rooftop equipment, taking into account positioning that avoids shading of solar equipment and maximization of continuous roof space;[34]
      iii. Sizing and/or provision of extra electrical panels to accommodate addition of an appropriately sized future solar energy system; [35] and
      iv. Provision of space for a solar energy system DC-AC inverter in the utility room or on an outside wall.
   D. The Department shall also consider including requirements for:
      i. Roof orientation and angle;[36]
      ii. Roof types that are compatible with a solar installation mounting strategy that will require minimal or no roof penetrations;[37] and
      iii. A conduit for wiring from roof to electric panel.
   E. To the extent necessary, the Department shall promulgate separate standards for residential and commercial construction.
4. Solar Hot Water Heater Readiness Standards

A. The Department shall promulgate regulations by [*date*] that establish minimum standards that must be met for new construction to accommodate a solar hot water heater.
B. In drafting the regulations, the Department shall take into account existing building code requirements and compliance costs. The Department shall also consult with scientists, engineers, professional societies, and government agencies with relevant expertise in solar hot water heating and building construction.
C. At a minimum, the Department shall include requirements for:
    i. Roof strength, with a requirement that roofing where solar hot water heating equipment could be placed be capable of supporting a minimum of 6 lb/ft^2;
    ii. Placement of non-solar-related rooftop equipment, taking into account positioning that avoids shading of solar equipment and maximization of continuous roof space;[38]
    iii. Provision of sufficient free space in utility rooms for solar hot water equipment at a minimum of one times the space required for the conventional water heating equipment;
    iv. Plumbing that will allow solar hot water piping to be attached to a conventional system without requiring the building water system to be de-pressurized;[39]
    v. Plan hot water distribution systems to accommodate SHW system integration;
    vi. On-demand water heaters that can operate in conjunction with a SHW system or implement distributed domestic hot water tanks that are capable of being integrated with a SHW system; and
    vii. Roof types that are compatible with a solar installation mounting strategy that will require minimal or no roof penetrations.[40]
D. The Department shall also consider including requirements for:
    i. Plumbing installed from the utility room to the roof of the building and
    ii. Metering of hot water load in commercial buildings.
E. To the extent necessary, the Department shall promulgate separate standards for residential and commercial construction.

5. Exemptions

A. Developers may seek an exemption from [*building code official or the Department*] from the solar system and/or solar hot water heater readiness requirements upon a sufficient showing that:
   i. Accommodation of a solar energy system and/or a solar hot water heater would be impractical due to poor solar resources at a site of new construction;
   ii. A substitute renewable energy system will be installed at the time of new construction; or
   iii. A substitute renewable hot water system will be installed at the time of new construction.

B. The Department shall issue regulations that define what constitutes a sufficiently poor solar resource such that a building should be exempted. The Department shall also consider publishing a list of acceptable software tools that can be utilized to determine the available solar resources for a building.

C. The Department shall publish regulations by [date] that clearly define the process for seeking an exemption.

6. Availability of Incentives
   Compliance with the provisions of this [Act] shall not impair a pre-existing building's eligibility for any incentives, rebates, credits, or other programs in existence to encourage development of renewable resources.
7. Enforcement
   A building permit for new construction may not be granted without a showing that the building complies with the requirements of this [*Act*].

APPENDIX C. SAMPLE OF SOLAR READY REQUIREMENTS OR PRACTICE BY JURISDICTION

1. Arizona

1.1. Tucson – Solar Design Requirement for Homes

1.1.1. Type of Technology

The requirement addresses photovoltaic (PV) and solar water heating systems.

1.1.2. Legislative Mandate/Option

Tucson adopted an ordinance in June 2008 that requires all new single-family homes and duplexes in Tucson to be solar ready. The ordinance requires all new homes either to have a PV and solar water heating system installed, or to have all the necessary hardware installed so that a system can easily be installed at a later date.

City of Tucson Energy Office
4004 S. Park Ave., Bldg. 1
Tucson, AZ 85714

Bruce Plenk
Phone: 520-791-5111
Email: bruce.plenk@tucsonaz.gov
Website: http://cms3.tucsonaz.gov/energy

Database of State Incentives for Renewables and Efficiency:
Arizona, City of Tucson – Solar Design Requirement for Homes
http://www.dsireusa.org/incentives/incentive.cfm?Incentive_Code=AZ26R&re=1&ee=1

1.1.3. Technical Requirements

To comply with this requirement, new homes must either have a complete solar water heating system installed or comply with one of two solar stub-out options. Option one requires the installation of two insulated pipes and a suitably sized conduit (for two pairs of monitoring and control wires) that run from the water heater area through the roof and are capped. Option two does not require the installation of pipes, but it does require the installation of a sleeve or conduit of sufficient size to hold the two insulated pipes and wires. With option two there must be a straight line from the water heater area to the roof. Both options will greatly cut down on the cost of installing a system at a later date. To comply with the PV requirement, a site plan must indicate the best roof space for locating the PV panels, and provide a roof structure designed for the additional collector weight. The site plan must also illustrate the best space available for accommodating PV equipment (meter, inverter, disconnect), and it should be adjacent to the electrical service panel or on a wall near the proposed location of the panels. There must also be a minimum 3,800-volt-ampere PV electrical load entry on the service load calculation, and an Electrical Panel Schedule with a 240-volt circuit breaker space labeled "reserved for photovoltaic."

1.1.4. Agency Responsible for Specifications

The ordinance was developed by a stakeholder group that included Technicians for Sustainability, the Tucson Association of Realtors, the Sierra Club, the Southern Arizona Homebuilders Association, architectural professionals, solar energy companies, and elements of the city government.

1.1.5. Exceptions

These requirements may be waived if it can be demonstrated to the building official that compliance is not practical due to shading, building orientation, construction constraints, or the configuration of the parcel of land.

2. California

2.1. California State Energy Code

2.1.1. Type of Technology

The requirement addresses PV and solar water heating systems.

California Energy Commission

Patrick Saxton
Phone: 916-651-0589
Email: psaxton@energy.state.ca.us

Database of State Incentives for Renewables and Efficiency:
California State Energy Code
http://www.dsireusa.org/incentives/incentive.cfm?Incentive_Code=CA
51R&re=1&ee=1

2.1.2. Legislative Mandate/Option

California Green Building Standards Code (GBSC) (Title 24, Part 11). On page 60 (page 69 of the .pdf), there are residential voluntary measures for "space for future solar installation" (A4.211.3) and "future access for solar system" (A4.211.4).

2.1.3. Technical Requirements

On page 127 (page 136 of the .pdf), there are nonresidential voluntary measures for "prewiring for future solar" (A5.211.4) and "off-grid prewiring for future solar" (A5.211.4.1). Local jurisdictions may make portions of the GBSC mandatory, but there are no specifics. To view the text of the local ordinances go to:www.energy.ca.gov/title24/2008standards/ordinances/. Note that these local ordinances are primarily meant to exceed the California Energy Code (Title 24, Part 6) but may also contain additional elements, such as those from the GBSC.

2.2. Marin County – Single-Family Dwelling Energy Efficiency Ordinance

2.2.1. Type of Technology

The requirement addresses PV and solar water heating systems.

2.2.2. Legislative Mandate/Option

Marin County Code 19.04.100.

Marin County Community Development Agency

3501 Civic Center Dr. #308
San Rafael, CA 94903-4157

Omar Pena
Phone: 415-507-2797
Website:
http://www.co.marin.ca.us/depts/CD/main/comdev/ADVANCE/ index.cfm

Database of State Incentives for Renewables and Efficiency:
Marin County – Green Building Requirements http://www.dsireusa.org/incentives/incentive.cfm?Incentive_Code=CA40R&re=1&ee=1

2.2.3. Technical Requirements

New commercial buildings over 5,000 square feet and all new residential construction must include plumbing and electrical conduit specifically designed to accommodate the easy installation of solar water heaters and PV systems in the future. These systems can be used to comply with the building code, but they are not required by the building code.

2.2.4. Agency Responsible for Specifications

The exact requirements for these solar ready additions will be adopted by the chief building inspector.

3. Colorado

3.1. Colorado Building Energy Code with Mandatory Solar Option

3.1.1. Type of Technology

The requirement addresses PV and solar water heating systems.

3.1.2. Legislative Mandate/Option

HB 1149, enacted in May 2009, requires builders of single-family homes to offer solar as a standard feature to all prospective homebuyers.

Governor's Energy Office
1580 Logan Street, Suite 100
Denver, CO 80203

James Dillon
Phone: 303-866-2100
Fax: 303-866-2930
E-Mail: jamil.dillon@state.co.us

Database of State Incentives for Renewables and Efficiency:
Colorado Building Energy Code with Mandatory Solar Option
http://www.dsireusa.org/incentives/incentive.cfm?Incentive_Code=CO36R&re=1&ee=1

3.1.3. Technical Requirements

Builders are required to give the buyer the option to either have a PV system or a solar water heating system installed on their new home, or to have all the necessary wiring and/or plumbing installed so they can easily add a solar system at a later date. The builder must also provide the buyer with a list, maintained by the Governor's Energy Office, of every solar installer in the area, so the buyer can obtain expert help in determining if their home's location is suitable for solar and what the estimated cost savings would be.

3.2. Boulder Green Points Building Program

City of Boulder, Department of Community Planning & Sustainability

Brian Holmes
Zoning Administrator
Phone: 303-441-3212
Fax: 303-441-3241
Email: HolmesB@bouldercolorado.gov

Database of State Incentives for Renewables and Efficiency:
City of Boulder – Green Points Building Program
http://www.dsireusa.org/incentives/incentive.cfm?Incentive_Code=CO06R&re=1&ee=1

3.2.1. Type of Technology

The requirement addresses PV and solar water heating systems.

3.2.2. Legislative Mandate/Option

There is no code or ordinance. But on page 127 (page 136 of the .pdf), there are nonresidential voluntary measures for "prewiring for future solar" (A5.211.4) and "off-grid prewiring for future solar" (A5.211.4.1).

3.3. Aspen & Pitkin County – Renewable Energy Mitigation Program

3.3.1. Type of Technology

The requirement addresses solar water heat, PV, and geothermal heat pumps.

3.3.1.1. Legislative Mandate/Option

Pitkin County does not require homes to be solar ready. It does, however, have a mandatory Efficient Building Program (EBP) that requires residences to score points by incorporating sustainable building measures in home construction. Solar ready measures are one of the ways the builder may choose to score points. The EBP is mandatory—a building permit cannot be issued without scoring enough points. To summarize, some homes in unincorporated Pitkin County are solar ready, but the percentage is very low.

Pitkin County

Tony Fusaro
Chief Building Official
Phone: 970-920-5101
Website: www.aspenpitkin.com

Database of State Incentives for Renewables and Efficiency:
Aspen & Pitkin County – Renewable Energy Mitigation Program
http://www.dsireusa.org/incentives/incentive.cfm?Incentive_Code=CO16R&re=1&ee=1

4. Guam

4.1. Guam – Solar Ready Residential Building Requirement

4.1.1. Type of Technology

The requirement addresses solar hot water heaters.

Guam Energy Office
Pacific Energy Resources Center
548 N. Marine Drive
Tamunung, GU 96911

Lorilee Chrisostomo
Phone: 617-646 4361
Fax: 671-649-1215 Email: lorilee@guamcell.net
Website: www.guamenergy.com

Database of State Incentives for Renewables and Efficiency:
Guam – Solar-Ready Residential Building Requirement
http://www.dsireusa.org/incentives/incentive.cfm?Incentive_Code=GU01R&re=1&ee=1

4.1.1.1. Legislative Mandate/Option

The Guam Energy Code became effective in October 2000.

4.1.2. Technical Requirements

The requirement mandates that piping stub-outs be provided for water heaters installed in low-rise residential buildings to enable the future installation of solar collectors.

5. New Jersey

5.1. New Jersey Solar Energy Option Requirement for Residential Developments

5.1.1. Type of Technology

Broad definition could include passive as well as active solar elements.

New Jersey Department of Community Affairs

101 South Broad St.
PO Box 802
Trenton, NJ 08625-0802

Darren Port
Division of Codes and Standards
Phone: 609-984-7607
Website: http://www.state.nj.us/dca/divisions/codes/index.html

Database of State Incentives for Renewables and Efficiency:

New Jersey Solar Energy Option Requirement for Residential Developments
http://www.dsireusa.org/incentives/incentive.cfm?Incentive_Code=NJ18R&re=1&ee=1

5.1.1.1. Legislative Mandate/Option

In March 2009, New Jersey enacted legislation designed to support the integration of solar energy systems into new residential developments. The law requires that, whenever "technically feasible," developers of residential developments with 25 or more dwelling units (i.e., single-family residences) offer to install or provide for the installation of a solar energy system on the unit during negotiations with a prospective purchaser. Developers are required to disclose this option in their advertising materials, which must include information on installation costs, environmental benefits, energy cost savings, and incentive programs for which the installation may qualify.

5.1.2. Technical Requirements

NJ//P.L. 2009, CHAPTER 33/3 stipulates that the solar energy system is to be installed in conformance with the manufacturer's specifications and in compliance with all applicable electrical and building code standards.

5.1.3. Agency Responsible for Specifications

The Board of Public Utilities shall adopt orders, rules, or regulations that provide for solar energy systems installed in accordance with the provisions of P.L. 2009, c.33.

6. Massachusetts

6.1. City of Boston

6.1.1. Type of Technology

The requirement addresses PV systems.

6.1.1.1. Legislative Mandate/Option

The City of Boston Department of Neighborhood Development (DND) coordinates the construction of affordable housing units within the city. DND has worked with the Solar Boston Initiative to incorporate solar into the department housing portfolio through the Green Affordable Housing Program (GAHP) as part of a $1.8 million grant from the Massachusetts Clean Energy Center. The DND has required that all future affordable housing development in the city be built to solar ready guidelines.

City of Boston

Andrew Belden
Department of Neighborhood Development
Email: Andy.Belden@mc-group.com

6.1.2. Technical Requirements

The solar ready guidelines, which were developed with the help of NREL, require new affordable housing developments to limit roof obstructions and to avoid roof designs that would complicate future solar installations. The solar ready standard has been in place since 2007 for all DND-developed affordable housing projects.

End Notes

[1] Economics and Statistics Administration. New Residential Construction Report. http://www.esa.doc.gov/sites/default/files/ei/documents/2011/January/new_residential_construction_december_2010.pdf. Accessed April 2011.

[2] L. Lisell, T. Tetreault, and A. Watson. *The Solar Ready Buildings Planning Guide*. NREL/TP-7A2-46078. Golden, CO: National Renewable Energy Laboratory, December 2009. Available online at http://www.nrel.gov/docs/fy10osti/46078.pdf.

[3] ENERGY STAR. http://www.energystar.gov/. See "Certification"/"Designed for EnergyStar." Accessed April 2011.

[4] LEED Homes Rating System. http://www.usgbc.org/DisplayPage.aspx?CategoryID=19. Accessed April 2011.

[5] Waier, P.R., ed. *Green Building Cost Data*. RSMeans, 1st Annual Edition, Norwell, MA: RSMeans, 2010.

[6] Waier, P.R., ed. *Facilities Maintenance & Repair Cost Data*. RSMeans, 18th Annual Edition, Norwell, MA: RSMeans, 2010.

[7] Waier, P.R., ed. *Green Building Cost Data*. RSMeans, 1st Annual Edition, Norwell, MA: RSMeans, 2010.

[8] New Jersey AB 1558. http://www.njleg.state.nj.us/2008/Bills/PL09/33_.PDF. Accessed April 2011.

[9] General Assembly of the State of Colorado: Section 1, Article 35.7 of title 38, Colorado Revised Statutes. http://www.leg.state.co.us/clics/clics2009a/csl.nsf/fsbillcont3/56DAD78B9D26BD5187257539006E7FF9?open&file=1149_enr.pdf. Accessed April 2011.

[10] Online Code and Environment Advocacy Network (Ocean). http://bcap-ocean.org/state-country/colorado. Accessed April 2011.

[11] General Assembly of the State of Colorado: Section 1, Article 35.7 of title 38, Colorado Revised Statutes. www.leg.state.co.us/clics/clics2009a/csl.nsf/fsbillcont3/56DAD78B9D26BD5187257539006E7FF9?open&file=1149_enr.pdf. Accessed April 2011.

[12] New Jersey AB 1558. www.njleg.state.nj.us/2008/Bills/PL09/33_.PDF. Accessed April 2011.

[13] NREL. In My Backyard. www.nrel.gov/eis/imby/. Accessed April 2011.

[14] SolOpt Optimization Tool, available at the Solar America Communities website. http://solaramericacommunities.energy.gov/Publications.aspx. Accessed April 2011.

[15] Pima County Regional Residential Rating System, Tucson, Ariz. http://www.pimaxpress.com/green. Accessed April 2011.

[16] Database of State Incentives for Renewables & Efficiency (DSIRE). Arizona: Solar Rights http://dsireusa.org/incentives/incentive.cfm?Incentive_Code=AZ07R&re=1&ee=1. Accessed April 2011.

[17] Colorado General Assembly. House Bill 08-120. www.leg.state.co.us/clics/clics2008a/csl.nsf/fsbillcont3/3F45E0C88F60DD39872573680059F7E2?open&file=1270_enr.pdf. Accessed April 2011.

[18] Colorado General Assembly. Digest of Bills. www.state.co.us/gov_dir/leg_dir/olls/digest 2008a/PROPERTY.htm#08-1270. Accessed April 2011.

[19] Database of State Incentives for Renewables & Efficiency (DSIRE). Georgia: Solar Easements http://dsireusa.org/incentives/incentive.cfm?Incentive_Code=GA01R&re=1&ee =1. Accessed April 2011.

[20] Database of State Incentives for Renewables & Efficiency (DSIRE). California: City of Sebastopol Solar Access http://dsireusa.org/incentives/incentive.cfm?Incentive_Code= CA33R&re=1&ee=1. Accessed April 2011.

[*] The Colorado law uses the term energy efficiency, which includes renewable energy like solar.

[21] Database of State Incentives for Renewables & Efficiency (DSIRE). See "State Incentives/ Policies for Solar." www.dsireusa.org/solar/.

[22] University of Wisconsin Center for Land Use Education. 2005. "Planning Implementation Tools: Density Bonus." www.uwsp.edu/cnr/landcenter/pdffiles/implementation/density bonus.pdf. Accessed April 2011.

[23] Peters, Katherine McIntire. "ENERGY STAR Program Comes Under Attack." March 30, 2010. http://www.govexec.com/dailyfed/0310/033010kp1.htm. Accessed April 2011.

[24] ENERGY STAR. "Apply for the ENERGY STAR for Your Buildings." http://www. energystar.gov/index.cfm?c=evaluate_performance.bus_portfoliomanager_intro. Accessed April 2011.

[25] ENERGY STAR. "Design to Achieve Energy Star." http://www.energystar.gov/index.cfm?c= cbd_guidebook.cbd_guidebook. Accessed April 2011.

[26] ENERGY STAR. "Renewable Energy Ready Homes (RERH)." http://www.energystar.gov/ index.cfm?c=rerh.rerh_index. Accessed December 2011.

[27] ENERGY STAR: ENERGY STAR for Homes Version 3 Guidelines. http://www.energystar. gov/index.cfm?c=bldrs_lenders_raters.nh_v3_guidelines. Accessed April 2011.

[28] U.S. Green Business Council. http://www.usgbc.org/DisplayPage.aspx?CMSPageID=283. Accessed April 2011.

[29] LEED Rating System Draft. November 2010. www.usgbc.org/ShowFile.aspx?DocumentID= 8186. Accessed April 2011 (This draft is no longer available to the public).

[30] U.S. Green Building Council. http://www.usgbc.org/. Accessed April 2011.

[31] City of Boston Department of Neighborhood Development (DND). www.cityofboston.gov/ dnd/d_green_housing.asp. Accessed April 2011.

[32] As technical assistance to the City of Boston, NREL designed solar ready guidelines that the city can use as an educational document to inform stakeholders. See Solar Ready Buildings Planning Guide, http://www.nrel.gov/docs/fy10osti/46078.pdf.

[33] Lunning Wende Associates, Inc. *Solar Ready Building Design Guidelines for the Twin Cities, Minnesota*. September 2010. www.state.mn.us/mn/externalDocs/Commerce/Solar_Ready_ Building_Design_Guidelines_020211042659_SolarBuildingDesignGuidelines.pdf. Accessed April 2011.

[34] To optimize future solar energy and hot water systems, NREL recommends that there be at least 75% continuous roof space, unobstructed by HVAC equipment, vents, or any other objects, and that all such equipment be located on the north end of the solar ready roof. www.nrel.gov/docs/fy10osti/46078.pdf. Accessed April 2011.

[35] To optimize future solar energy and hot water systems, NREL recommends that residential electrical panels have an excess of 30% capacity and that space for an additional electric panel be provided in commercial buildings. www.nrel.gov/docs/fy10osti/46078.pdf. Accessed April 2011.

[36] To optimize future solar energy and hot water systems, NREL recommends that all buildings be oriented with a sky view within 15 feet of due south or be flat in order to be considered solar ready. www.nrel.gov/docs/fy10osti/46078.pdf. Accessed April 2011.

[37] To optimize future solar energy and hot water systems, for flat roofs NREL recommends fully adhered thermoplastic olefin or polyolefin (TPO) roofing. For slanted roofs, NREL

recommends a structurally fastened standing seam roof. www.nrel.gov/docs/fy10osti/46078.pdf. Accessed April 2011.

[38] NREL recommends that there be at least 75% continuous roof space, unobstructed by HVAC equipment, vents, or any other objects, and that all such equipment be located on the north end of the solar ready roof.

[39] NREL recommends requiring that T-joints and shut-off valves be required to accomplish this purpose.

[40] To optimize future solar energy and hot water systems, for flat roofs NREL recommends fully adhered thermoplastic olefin or polyolefin (TPO) roofing. For slanted roofs, NREL recommends a structurally fastened standing seam roof. www.nrel.gov/docs/fy10osti/46078.pdf. Accessed April 2011.

In: Solar Ready Buildings and Homes
Editors: W. I. Cook and N. M. James
ISBN 978-1-62081-573-1

Chapter 2

SOLAR READY BUILDINGS PLANNING GUIDE*

L. Lisell, T. Tetreault and A. Watson

INTRODUCTION

This document identifies the important aspects of building design and construction to enable installation of solar photovoltaic and heating systems at some time after the building is constructed. This document addresses photovoltaic (PV), solar hot water (ST), and solar ventilation preheat (SVP) systems.

Often, the high initial cost of PV, ST, and SVP systems prevents them from being included in new construction. However, with better incentives, technological improvements, and rising conventional power prices, energy from solar sources will become more cost competitive. New construction that is solar ready will be in a position to take advantage of an environment more favorable to renewable energy. Without the forethought to make buildings solar ready, solar installation may not be technically possible or the added costs of making infrastructure changes may make solar applications economically prohibitive.

Planning for the eventual installation of a solar system when designing a building can significantly improve the economics of the investment.

* This is an edited, reformatted and augmented version of the Technical Report NREL/TP-7A2-46078, U.S. Department of Energy, dated December 2009.

Installation efficiency can be maximized and costs minimized by understanding these systems' requirements and accounting for them during the design and construction of the building.

System performance is another significant contributing factor in the economics of PV, ST, and SVP systems. Similar to installation efficiency, system performance can be maximized by planning ahead during the design of the building.

The following guidelines indentify important system requirements for PV, ST, and SVP systems. Attention to these guidelines in the development of building codes, building design, or any building- or community-related regulations could significantly improve the performance and minimize the cost of solar systems.

About This Guide

This document should be used as a guide to influence the design of new solar ready buildings. It outlines the scope of consideration that will minimize solar installation costs and maximize solar production potential when solar technologies are applied.

The guide is divided into two sections. A checklist provides quick reference and outlines the areas that need to be considered. The checklist references a second, descriptive section that provides detail and background information to the user.

This document was designed to be concise and hence accessible to users. When implementing the strategies outlined here, other resources must be consulted. The National Electric Code and local jurisdictions have specific requirements that apply to solar and must be considered. Local solar installers will be valuable resources at this early stage.

This guide was written with the focus on new construction. While many of the considerations are similar for retrofit or renovated buildings, there are other considerations to be made for retrofit projects that are not outlined here.

Solar Ready Checklist

This guide is laid out in checklist form, with further information on each checklist item in the document below. The checklist will serve as a table of contents for the document as well as a quick reference tool for use when building plans are being developed.

General Solar Guidelines

Avoid shading from trees, buildings, etc. (especially during peak sunlight hours). (More Info – Avoid Shading)	☐
Check the zoning laws for the proposed site to ensure that future, neighboring construction will not cast shade on the array. (More Info – Zoning Laws)	☐
Determine where a future solar array might be placed. (More Info – Solar Collector Placement)	☐
If the roof is sloped, the south-facing section will optimize the system performance; keep the south-facing section obstruction-free if possible. (More Info – Site Orientation)	☐
Minimize rooftop equipment to maximize available open area for solar collector placement. (More Info – Roof Obstructions)	☐
The type of roof installed can greatly affect the cost of installing solar later. (More Info- Roof Types)	☐
The roof must be capable of carrying the load of the solar equipment. (PV – between 3 and 6 lb/ft2) (ST – between 2 and 5.5 lb/ft2) (More Info – Roof Load)	☐
The wind loads on rooftop solar equipment must be analyzed in order to ensure that the roof structure is sufficient. See the American Society of Civil Engineers (ASCE) international building code 7-05 for the method of how to calculate these loads. (More Info – Wind Load)	☐
Record Roof Specifications on Drawings. (More Info – Record Roof Specs)	☐
Add additional safety equipment for solar equipment access and installation. (More Info – Safety Equipment)	☐
Decide where the solar panels will be mounted, and consider the different mounting strategies available. If the panels will be mounted on the roof with penetrating hardware, consider installing the mounting hardware at the time the roof is installed and use flashings for every penetration. (More Info – Panel Mounting)	☐
If the collectors will be placed on the roof, check if the roof installation carries a warranty. Determine if the roof warranty contract has terms involving solar installation. (More Info – Roof Warranty)	☐
Make sure all equipment is in compliance with the current version of the National Electrical Code. (More Info – Verify System Compliance)	☐

Additional Guidelines for Solar Thermal (ST) Systems

Determine the hot water load and water temperature requirement to size all of the system components. Residential hot water usage can be calculated using the "Building America Research Benchmark Definition" by R. Hendron. (More Info – Sizing the System (ST))	☐
Identify location for controllers, heat storage system, shutoff valves, and other equipment. (More Info – Equipment Space (ST))	☐
Plan for all the necessary plumbing and provide pathways for water lines to link the solar collector, the heat storage system, and the rest of the building's hot water system. Also, be sure that the water distribution system and mixing valves are laid out to maximize the ST system performance. (More Info – Plumbing (ST))	☐

Additional Guidelines for Solar Ventilation Preheat (SVP) Systems

Minimize the number of penetrations (windows, doors, etc.) through the south-facing wall. The transpired collector can be installed around penetrations but excessive irregularities in the geometry can limit the system performance. (More Info – Southern Wall Exposure (SVP))	☐
Provide direct access from transpired collector location on south wall to the outside air (OA) intake duct of the air handling unit. (More Info - Duct Work Connection Access (SVP))	☐
Install a knockout and bypass damper in the first stage of the air handling unit to accommodate ductwork from the transpired collector. (More Info – Duct Work Connection Access (SVP))	☐
Air handling unit fan should be sized to accommodate additional pressure drop from the transpired collector. (More Info – Additional HVAC Equipment (SVP))	☐
Design control system to accommodate additional control points to control dampers. (More Info – Additional HVAC Equipment (SVP))	☐
Design sufficient space between point sources of emissions (parking lots, delivery zones, etc.) and the transpired collector. (More Info – Southern Wall Exposure (SVP))	☐
Design south wall to fire code standards to accommodate the installation of the transpired collector. (More Info – Southern Wall Exposure (SVP))	☐

Additional Guidelines for Photovoltaic (PV) Systems

Identify electrical panel location for convenient PV system inter-connections, and keep space available in the electrical panel for a PV circuit breaker. (More Info – Electrical Panel Location (PV))	☐
Specify panel capacity sufficient to accommodate the total power coming into the building (proposed PV system size power generation plus size of breaker protecting main panel). NEC allows for the sum of these two sources of power to be 20% greater than the panel rating. Consult the local authority having jurisdiction. (More Info – Electrical Panel Location (PV))	☐
Lay out the locations for the inverter and the balance of system (BOS) components. (More Info – Wiring Schematic (PV))	☐
Identify the inter-connection restrictions for the location of the building site that apply to grid-tied PV systems. Begin by reviewing interconnection standards at http://www.dsireusa.org/. (More Info – Grid Inter-connection (PV))	☐
Run electrical conduit from the solar collector location to the electrical panel and other electrical components. (More Info – Wiring Schematic (PV))	☐
Consider any special load needs (i.e., uninterrupted power supply) and consider whether storage is needed. (More Info – Consider Special Loads (PV))	☐
PV panels are much more sensitive to shading than ST panels. Avoid shading as much as possible. Due to the individual modules of a PV panel being connected in series, even a narrow strip of shading (lightning rods, antennas, etc.) can limit the current of the entire array. (More Info – Avoid Shading)	☐
Find out what the energy production of the proposed system will be using the PVWatts calculation tool, and adjust the system size as needed: http://rredc.nrel.gov/solar/codes_algs/PVWATTS/version1/. (More Info – Sizing the System (PV))	☐

DETAILED SOLAR READY BUILDING GUIDELINES

General Solar Guidelines

Requirements for buildings to be ready for any type of solar system are detailed below.

Avoid shading

Shading will have an adverse effect on all the solar technologies discussed here and should be avoided. Use a sun path calculator, such as the Solar Pathfinder™, to assess shading at a particular location by analyzing the sky view where the solar panel will be located. Since the solar application may not be installed for several years, landscaping and future construction should be planned so as to avoid adversely affecting the solar resource.

PV panel performance is exceptionally susceptible to shading. When shade falls on a PV panel, that portion of the panel is no longer able to collect the high-energy beam radiation from the sun. If that shading happens during the peak hours of operation (10 a.m. – 2 p.m.), the production of the panel can be greatly reduced. A PV panel is made up of many individual cells that all produce a small amount of current and voltage. These individual cells are connected in series to produce a larger voltage.

Figure 1 shows how a typical panel is laid out. If enough of the individual cells are shaded, the voltage will drop below the lower voltage limit of the panel and the entire panel will quit producing electricity.

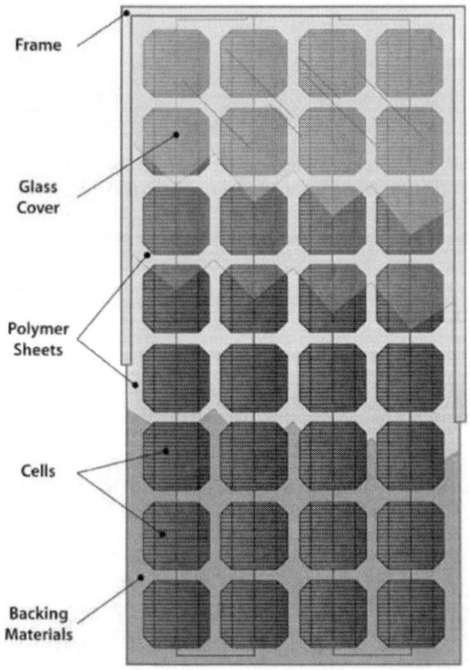

Figure 1. Typical PV panel construction.

Zoning Laws

It is important to know and understand the local zoning laws and how they impact the installation of solar panels and the preservation of access to the solar resource. Properties located within historical districts or covered by a homeowners association may be subject to additional restrictions. There are some local governments that have instituted laws that protect the solar access of existing PV arrays; these laws are commonly referred to as solar easements. In most places, zoning laws are only applied to residential buildings. These and other laws can have a significant impact on the design and development of solar ready buildings. Speaking with a local official who understands local zoning laws and solar easements can save time by eliminating zone violations early on in the process.

Solar Collector Placement

In many cases, the roof is the best location for a PV or ST system. This is a convenient location because it is out of the way and usually un-shaded. In addition to roof locations, there are many other areas that also make good collector locations. Some of these include parking shade structures, large plots of land, and building integrated systems. When considering a solar system, all possible locations should be considered.

Site Orientation

Before ground is broken on a new construction project, consider whether the site layout can be improved to better incorporate a solar system. In most cases, the layout of a site can determine whether a solar system is feasible or not. See Figure 2 for two example scenarios where site layout greatly changes solar potential. In Site 1, a large amount of open space is available north of the buildings, but because of shading from the buildings, the area will be shaded for a large part of the year. In Site 2, the buildings and the parking lot were shifted to the north side of the site. This left the open space on the south side of the site, where shading from the buildings will not fall on the solar panels. By placing the buildings on a site with solar resource and shading in mind, the area available for solar panels can be greatly increased.

If a building is designed with sloped roofs, it is best to orient the roof to maximize the roof area facing south (north-facing in the southern hemisphere). Placement on the south-facing section will ensure that the sunlight will strike the solar collector at a more optimal angle than it would if the collectors are placed on the east-, west-, or north-facing roof sections. The solar panels should be mounted on the south area of the roof and the mechanical equipment and vents should be located on the north area of the roof. Most solar collectors are mounted with a panel tilt angle, relative to horizontal, that is approximately equal to the site's latitude for maximum annual energy collection. Large systems with multiple rows have row spacing so each row gets full sun at 10 a.m. on the winter solstice.

Roof Obstructions

It is important to maximize the size of unobstructed roof top area, and large contiguous areas are preferable. Minimize the rooftop equipment if possible. If mechanical equipment must be mounted on the roof it is best to designate a confined area for this equipment to allow more space for a PV or ST array. When selecting the confined equipment area, select a place that will not shade the roof area intended for the solar array. Rooftop equipment includes small obstructions such as rooftop vents, which should also be minimized.

Roof Types

The type of roof installed when a building is built can affect installation costs when solar is pursued on the building. The type, quality, and warranty of the roof can all play a part in determining the ease of a solar install. Solar PV panels often have a 25-year warranty. It is important to install a roof that will last at least as long. Also, the type of roof chosen can determine whether roof penetrations will have to be made, which may void the roof warranty.

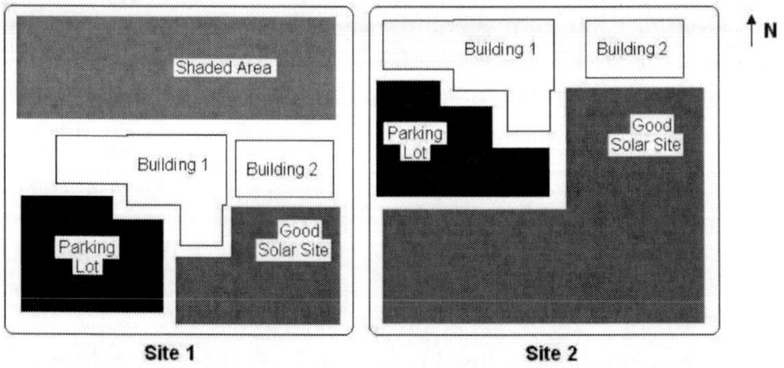

Figure 2. Usable solar area.

Flat Roofs

The best roof for a flat application is a fully adhered thermoplastic olefin or polyolefin (TPO) membrane roof. This roof is often an excellent choice for commercial applications, being both cost effective and more environmentally friendly than some other options. Other membrane roofs can also work well with solar such as ethylene propylene diene monomer (EPDM) or polyvinyl chloride (PVC). It is important to avoid river rock ballasted membrane roofs. Removing the ballast to install solar panels can be tedious and costly.

Sloped Roofs

A standing seam roof is often the best option if solar is to be pursued at a later date. Solar panels can be attached to standing seam roofs without penetrating the roof. The standing seam roof itself must be attached to the structure well enough to withstand the additional loads of a solar array. Any shingled roof will require roof penetrations in order to attach solar racking. A tile roof poses challenges for installing a solar array and should be avoided if solar is planned for the structure.

Roof Load

If the solar system is to be located on the roof of a building, roof structures must be designed to accommodate the additional dead loads (static load) and live loads (dynamic load) of the PV and/or ST system. Figure 3 illustrates the difference between live loads and dead loads. In the figure, the red arrows refer to the dead loads, and the blue arrows refer to the live loads. For a PV system, the solar panels and racking will add approximately 3 pounds for each square foot of collector area. If ballasted mounting is used, this number can be significantly higher ranging from 4-6 pounds per square foot of collector area. For a ST system, the solar panels will add different amounts of weight depending on the system type. For information on the selection of a ST system type visit the Solar Rating and Certification Corporation (SRCC) website at www.solar-rating.org. The following are examples from SRCC.

- Unglazed Solar Pool Heater – approx. 2.5 lb/ft2 of collector area[1]
- Glazed Flat Plate Collector – approx. 5.5 lb/ft2 of collector area[1]
- Evacuated Tube Collector – approx. 5.5 lb/ft2 of collector area.[1]

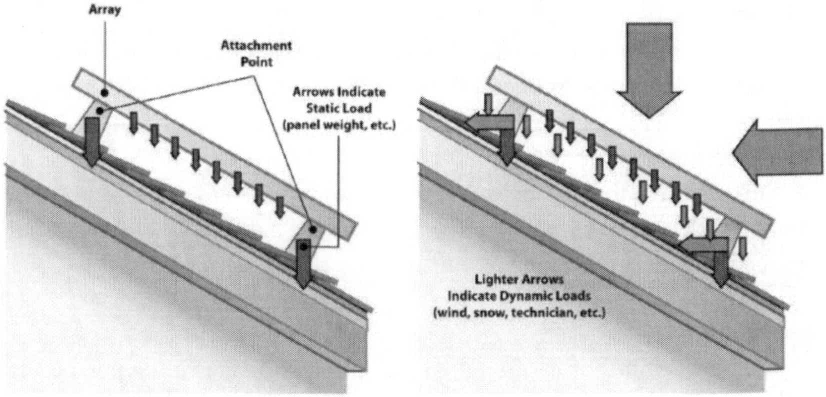

Figure 3. Live loads vs. dead loads.

Some ST systems have storage capacity coupled with the collectors. For an integrated collector, the system weight is highly dependent on size of the storage tank. If the collector can hold 80 gallons of water, the increase in weight is nearly 700 pounds. In order to design the roof to accommodate the added weight of an integrated collector – storage system, the system should be sized and the weight of the system calculated. A good resource that can be used to find information on the performance and exact weight of various ST collectors is: http://www.solar-rating.org/ratings/OG100DIRECTORIES/OG100DIRFULL.pdf.

In addition to the increased static load, consideration should be given to extraneous conditions that could be compounded due to a solar array (i.e., drifting snow, slow water runoff). In some cases this additional load can be many times the static load of a typical solar panel array. As mentioned above, when designing a building, it is critically important to be aware of the additional static and dynamic loads that could be imposed on the roof structure. The American Society of Civil Engineers (ASCE) international building code's chapter 7 can be used to calculate potential snow loads. There is more on loading in the next section.

Wind Load

Roof structures should be designed to accommodate the dynamic loads resulting from wind incident upon the solar collectors. The wind load is dependent upon the building location, wind conditions, collector orientation, height exposure category, topography of the surroundings, and the roof zone placement of the solar collectors. If the collectors are flush mounted, the wind load will be comparable to the wind load occurring on the roof. If the collectors are rack mounted, the wind load can be increased significantly. Figure 4 shows the additional shear and normal forces from wind loading, but there can also be a significant bending moment that should be considered. This bending moment is normally not of significant concern. However, where the flat roof trusses run parallel with the rows of solar collectors and the mounting rack is designed and anchored in a specific way, a significant bending moment can be exerted on the roof trusses. When designing a building, it is important to be aware of these potential additional loads to ensure that the roof structure can comfortably handle all of the eventual loading conditions. Most solar panel mounting manufacturers include loading factor worksheets with their mounting hardware that can be used to calculate loads. Another resource to consult is the American Society of Civil Engineers (ASCE) international building code 7-05. Wind loading, snow loading, and other important factors are laid out in this document as well as methods to calculate these factors. Upon completing the calculations, a local solar installer or local building official should be contacted to verify the results and check the building code.

Record Roof Specifications

If a roof has been designed to withstand additional loading from a solar array, it is crucial that this information is recorded so it is not lost to future solar developers. An ideal place to record this information is on the Drawing Package's Code Sheet.

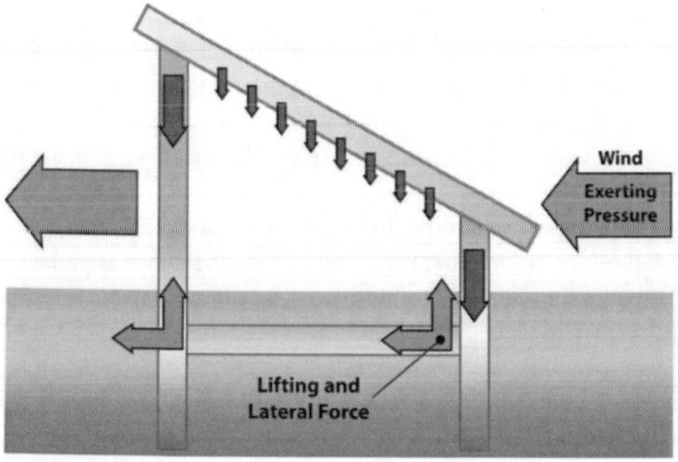

Figure 4. Wind loading.

Safety Equipment

Regardless of the location of the solar system, always consider appropriate safety measures. Maintenance, repairs, or inspections of the solar system require that a safety plan be developed. If the system is on a sloped roof, the building may need to be outfitted with a harness connection point. Additionally, fire code may require space around the edge of the roof where panels cannot be installed to allow safe access to all areas of the array. If the system is on the ground, a fence may need to be constructed around the system.

There are no national requirements for lightning protection of solar systems, but some organizations and state and local governments have internal requirements that should be considered during the design and construction of the building.

Panel Mounting

There are several strategies available for mounting solar panels. If the proper wind and loading analysis has been done on a solar system, the appropriate mounts can be chosen and installed at the time of the building construction. There are different types of mounts available depending on whether the panels are mounted on the ground, a flat roof, a slanted roof, a wall, a pole, or elsewhere.

For flat surfaces, ballasted systems are available. Ballasted systems are racks that hold solar panels in place by adding a sufficient amount of weight to the frame of the rack. An example of a ballasted system can be seen in Figure 5. These systems save time and possible roof damage by avoiding penetrations. They also require significant additional weight on roofs and can add labor costs for the temporary removal of the system if the roof ever needs to be resurfaced. Slanted roofs can be direct mounted, rack mounted or standoff mounted. Ground mounted systems have a variety of mounting options available.

Before choosing one mounting option, explore all of the mounting options available for the specific site and roof configuration. Each system has pros and cons associated with it; the best mounting option often varies from site to site.

If you find that roof-penetrating mounts are best suited for the application, consider installing the roof penetrating mounts at the time of the roof installation. Once a building is finished, it can be costly to install anchoring mounts through the roof into the structural supports. When selecting penetrating mounts, verify that there is sufficient penetration into structural supports for the loading specifications (wind load, snow load, dead load, etc.).

It is also important to consider the impact of roof penetrations on the insulating value of the roof. Penetrating mounts can create a thermal bridge between the interior and exterior space and can cause humid interior air (especially in indoor pools) to condense and drip from cold roof penetrations. There are ways to mitigate this issue but it is important to be aware of the impacts of roof penetrations on the building system.

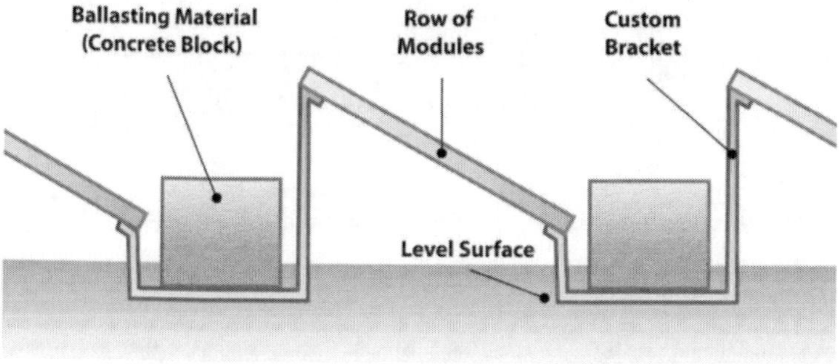

Figure 5. Ballasted mounting.

Figure 6. Penetrating mounts.

For every roof penetration, use flashing or flashed standoffs and appropriate sealing to ensure that the penetrations will not cause roof leaks. Consult a professional roofer for insight on the best way to seal specific roof types and verify that no roofing codes have been violated. Various penetrating mounts can be seen in Figure 6.

Another thing to consider for roof-mounted systems is the albedo rating of the roof. If the roof is highly reflective, the benefit is two-fold. The roof will reflect more radiation on to the solar collectors, and the roof will stay cooler due to less solar radiation being absorbed. If the temperature of the roof is lower, the PV panels will be cooler and will operate at higher efficiencies.

Roof Warranty

Included in most roof warranty contracts are certain terms and conditions that must be followed in order for the warranty to be valid. In some cases, a certain style of roof mount must be used in order to comply with a roof warranty contract. Whether it means using a ballasted mount, or installing a certain style of flashing at each penetration, it is important to address roof warranty questions during the design and development of the project.

Verify System Compliance with the National Electric Code

When the electrical schematic is laid out for the solar system, verify that all wiring and grounding requirements follow the latest version of the National Electric Code (NEC). The NEC must be followed for both PV and ST systems, in addition to the state electric code. NEC Section 690, "Solar Photovoltaic Systems," was laid out specifically for PV installations and outlines the necessary requirements for installing the electrical equipment for a PV system to code.

Solar Thermal (ST) Systems

Additional requirements for buildings to be ready for solar thermal systems are detailed below.

Sizing the System (ST)

The size of an ST system should be found by estimating the building hot water usage, water temperature requirements, and hot water usage patterns.

The "Building America Research Benchmark Definition" by R. Hendron is a good source of information for determining hot water usage of residential buildings. The document in PDF form can be found at http://www.nrel.gov/docs/fy08osti/42662.pdf. The information regarding hot water loads is located on page 11. The size of the system is useful for many structural calculations and also for mounting requirements. In order to correctly prepare for an ST system, the size of the system must be known. A general rule of thumb for cost effective ST applications is for a system to collect one-half to three-fourths of the annual thermal demands of a building. This solar fraction is dependent on the load profile and storage capacity of the system. A proper solar fraction helps prevent stagnation during the summer months in closed loop systems.

Equipment Space (ST)

An ST system requires space for hot water storage, plumbing, mixing valves, controllers, and other required equipment. The ST system should be selected and laid out in order to provide the dimensions and space requirements necessary for the system. Depending on the size of the system, the hot water storage can be a significant point load and should be addressed in the design of the flooring. A general rule of thumb is 1-2 gallons of storage is required for every square foot of collector. Included in this space is the clearance required by building and plumbing code. Access to electricity to power the ST system equipment must also be considered. There are a wide variety of ST system styles, and each one requires different plumbing and equipment. Some of the systems will require a large equipment space, and others will require very little. Listed below are five system schematics showing the main components. The mixing valve is an important component that protects against the hot water supply getting too hot and should be incorporated in every ST system. If the ST system required for a building is large, consider developing a plan for installing the various system components to ensure that there are no problems fitting the equipment into the building through openings (doorways, windows, etc.). Also, to make the eventual installation of the ST system easier, it is recommended to include strategically placed "T joints" and shut-off valves in the plumbing of the hot water system. See Figure 7 below for an illustration of a typical plumbing schematic.

Figure 8 is an example of a passive direct thermosyphon system. Water moves through this collector by convection only. The collector and the storage tank are mounted together, with the storage tank mounted just above the collector.

Figure 9 is a passive direct integral collector storage (ICS) system. Water moves through this collector by convection only, and the collector holds a large volume of water, eliminating the need for a storage tank. This system is self-contained and needs only to be plumbed into the building water system.

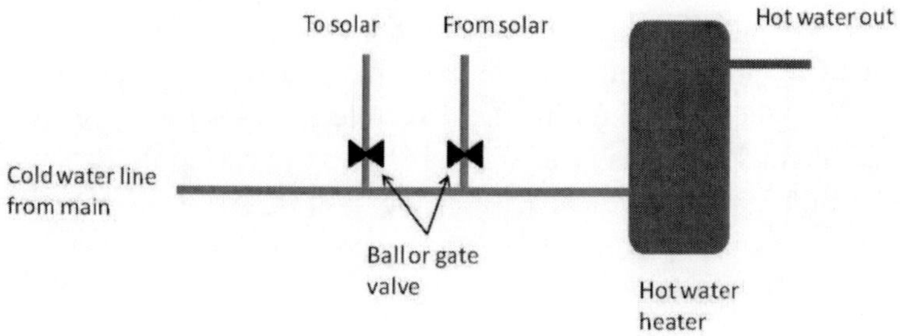

Figure 7. Plumbing schematic for eventual installation of ST system.

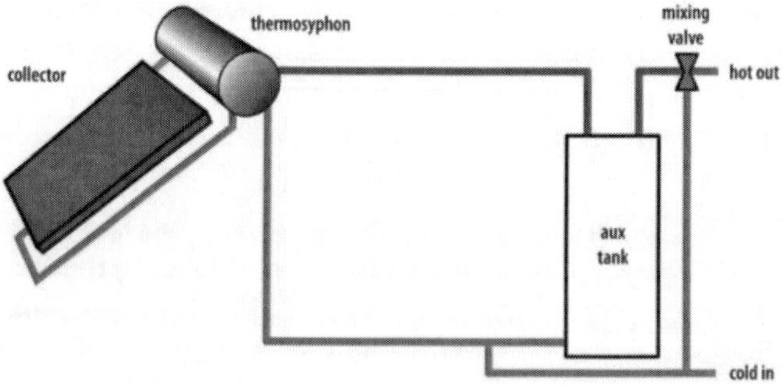

Figure 8. Thermosyphon system.

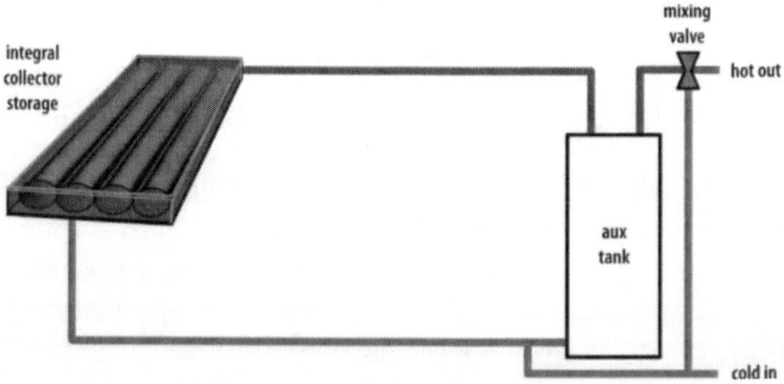

Figure 9. ICS system.

Figure 10 is an active direct system. Water is pumped around this system when the controller detects that the collector is hotter than the preheat tank. This system requires space for a preheat tank, the controller, the pump, and the additional plumbing.

Figure 11 is an active indirect system. The heating fluid and the water are pumped through the heat exchanger when the temperature of the collector is greater than the solar preheat tank. This system requires equipment space for the preheat tank, the pumps, the controller, the heat exchanger, and the additional plumbing.

Figure 12 is an active indirect system as well. The heating fluid drains out of the collector when the pump is not pumping. This system requires equipment space for the preheat tank, the pumps, the controller, the heat exchanger, the drainback tank, and all the additional plumbing.

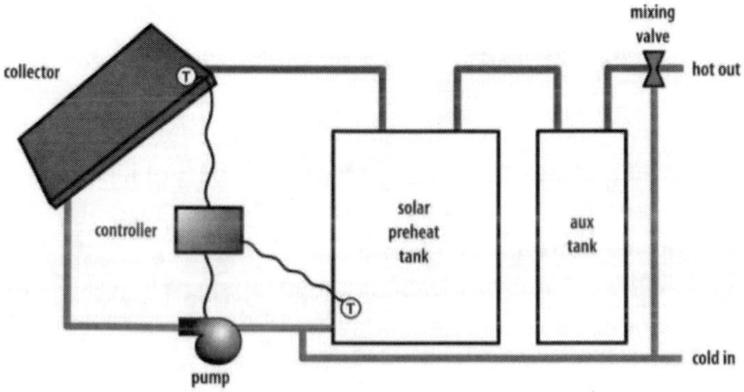

Figure 10. Direct system.

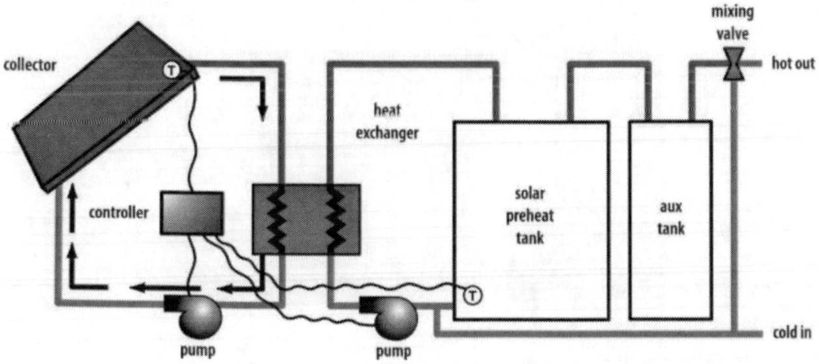

Figure 11. Indirect system.

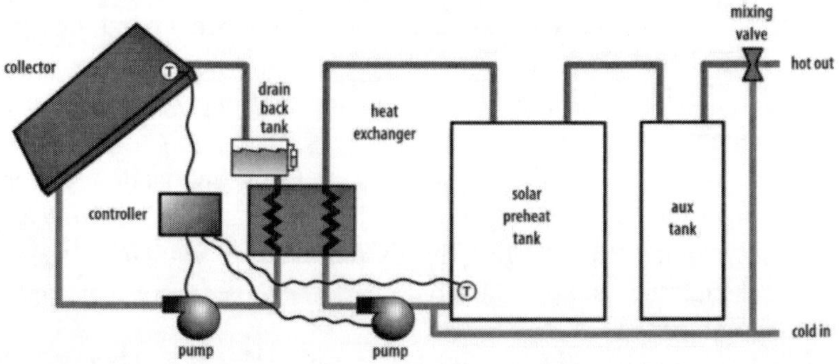

Figure 12. Drainback system.

Plumbing (ST)

The plumbing schematic for an ST system should be laid out at the time of building design. To enable ease of installation, plumbing from the roof to the equipment room should be installed when the building is first constructed. Any plumbing required for the system that goes through walls, floors, or other permanent structures should have the pathways planned out. If possible, there should be plumbing chases so that the piping can be installed quickly. Any plumbing that is installed should comply with all plumbing codes and plastic piping should *not* be used for ST applications. It is also recommended to use high temperature insulation on the ST system piping. The easily accessible plumbing can be done at the time of the system installation. Figure 13 shows some of the plumbing that must be installed along with an ST system.

Figure 13. ST system plumbing.

Solar Ventilation Preheat (SVP) Systems

PV and ST systems are applicable to most residential and commercial buildings, whereas SVP systems are only applicable to buildings with 100% outside air or otherwise high ventilation requirements. Additionally, the building should have a relatively long heating season, as solar ventilation preheating works best in moderate to cold and sunny climates. Examples of potential candidate buildings include: industrial buildings, warehouses (especially those that house materials requiring continuous ventilation), laboratory facilities, gymnasiums, and aircraft hangars. Additional requirements necessary for buildings to be ready for a Solar Ventilation Preheat system are detailed below.

Solar Ventilation Preheat System Overview

The solar ventilation preheat system works by heating outside air with a south-facing solar collector – a dark-colored wall made of sheet metal and perforated with tiny holes. Outdoor air is drawn through the holes and heated as it absorbs the wall's warmth. The warm air rises in the space between the solar wall and the building wall and is moved into the air-duct system, usually by means of a fan, to heat the building. Any additional heating needed at night or on a cloudy day is supplied by the building's conventional heating system. During the summer months, intake air bypasses the solar collector, preventing the system from supplying preheated ventilation air into the building. Figure 14 shows a diagram of a solar ventilation preheat system.

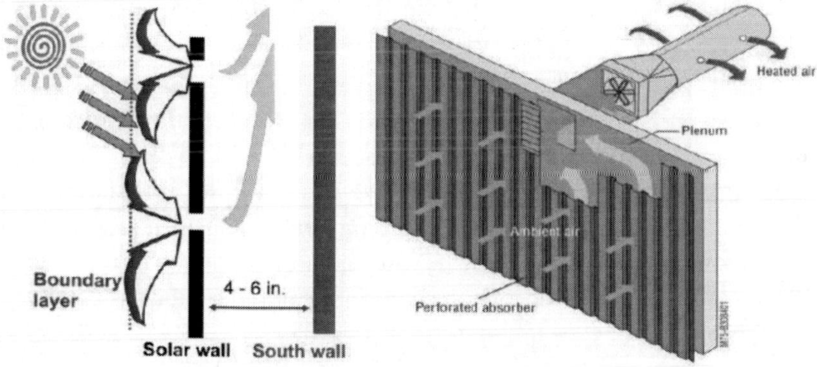

Figure 14. Solar ventilation preheat technology diagram.

Southern Wall Exposure (SVP)

Minimize the number of penetrations (windows, doors, vents, etc.) through the south-facing wall where the system could potentially be placed. Some penetrations can be worked around if necessary, but excessive irregularities in the geometry can limit the system performance. The system needs sunlight to strike the collector directly for optimal performance, so walls that are curved or have surfaces that are not due south will limit the effectiveness of the system. The mounting area on the wall must also be fire resistant. When sun is striking the collector and there is no heating load, the collector can get extremely hot. In order to eliminate the risk of fire, the wall must be designed to fire code standards. Keeping in mind that the collector is pre-heating the air entering the building, specify that there is sufficient space between the collector and emission sources. Emission sources to consider are exhaust vents, loading docks, parking lots, on site generators, etc.

Duct Work Connection Access (SVP)

To allow for the streamline tie-in of the SVP collector to the HVAC system, provide direct access from the potential collector location on the south wall of the building to the outside air (OA) intake of the air-handling unit. There should be a knockout in the OA duct to help facilitate the quick installation of the SVP system. The knockout will reduce the extra costs that could be accrued from installing extra ductwork, and also will ensure that the air does not cool significantly before reaching the heating system. In addition, there should be a bypass damper to allow the SVP collector to be bypassed while the building cooling system operates.

Additional HVAC Equipment (SVP)

The resistance of air through the SVP collector is larger than the resistance of air through an outside air duct, thus increasing the pressure drop from the outside to the air-handling unit. This increased pressure drop requires a larger fan to move the air. When specifying the size of fan for the air-handling unit, be sure that the fan is large enough to accommodate the increased pressure drop. In addition, specify that the HVAC control system has a sufficient number of control points to operate the dampers required for the system.

Photovoltaic (PV) Systems

Additional requirements necessary for a building to be ready for a PV solar system are detailed below.

Electrical Panel Location (PV)

Because the PV system needs to be connected to the electrical system of the building, the electrical panel should be in a convenient location to connect to the PV array. The electrical panel must have sufficiently large amperage rating to accommodate the PV energy as well as grid energy. By Section 690 of the NEC, the sum of the ratings of over current protection devices in all circuits supplying power to an electrical panel must not exceed 120% of the busbar rating. When the electrical panel for the building is selected, verify that the total energy coming into the building (proposed PV system size energy generation plus grid energy) does not exceed 120% of the panel rating. There must also be space available in the electrical panel for a PV circuit breaker. Near the electrical panel there should also be a convenient location for the inverter and balance of system (BOS) components, with a sufficient amount of clearance in front of the components to comply with NEC. Some of the electrical components can be located outside if necessary, but exposure to heat and the elements will reduce the performance of the components over time in some cases.

Grid Inter-connection (PV)

The grid inter-connection rules are different in every state, and sometimes even vary by city. Location-specific laws should be researched. The connection requirements vary depending on what type of grid the building is connected to and what the serving utility is. The serving utility representative should be contacted to determine what rules apply to the site. If the serving utility does not allow grid-interconnection, the possibility of implementing a solar array should not be eliminated. Each year more utilities are offering grid connection for solar arrays, and it is likely that it will be offered by all utilities in the near future. The rules and regulations associated with the grid inter-connection can be very specific, so it is important that the rules are fully understood. This process can be very time consuming, so it is beneficial to start the process well before the system is installed. A good place to get information on grid inter-connection and the various metering rules is: http://www.dsireusa.org/. Figure 15 shows two metering configurations that are widely used. These illustrate the inter-connection configuration known as "net metering". Net Metering is an electricity policy that allows a site with a PV system to send excess electricity to the grid and receive compensation. The compensation scheme is dependent on the servicing utility.

Wiring Schematic (PV)

The PV system will require conduit to go from the PV array to various electrical components. Figure 16 shows a PV system with an array and three of the many possible BOS components that must be connected to the array. Making sure there is adequate space for the BOS near the electrical panel is an easy step that can be taken during the building design phase and that will streamline installation at a later time.

An important consideration when designing a solar ready building is the conduit layout. Metallic conduit must run from the PV combiner box on the roof to the BOS space near the electrical panel. Ideally this conduit run is kept short to minimize voltage drop, and is planned for ahead of time for ease of installation. Running conduit through floors, ceilings, and walls during the initial construction will be much easier than doing it after the building is complete. Using an approximation of the eventual wiring schematic, conduit going through permanent structures should be installed. See the NEC for details on conduit and wiring requirements and verify that everything installed gets inspected for compliance with the latest NEC and the state electric and building codes.

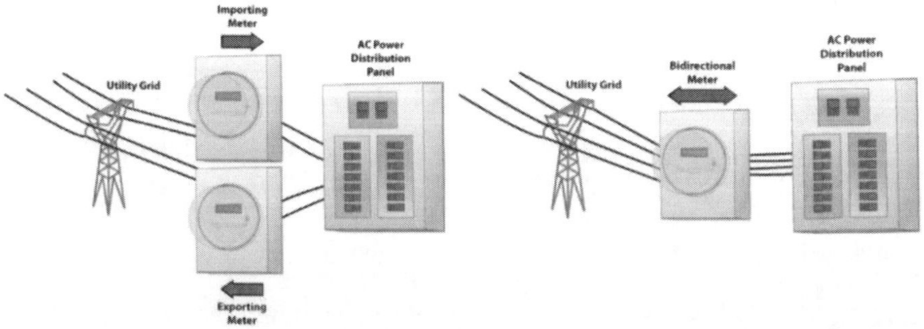

Figure 15. Dual and bi-directional metering configurations.

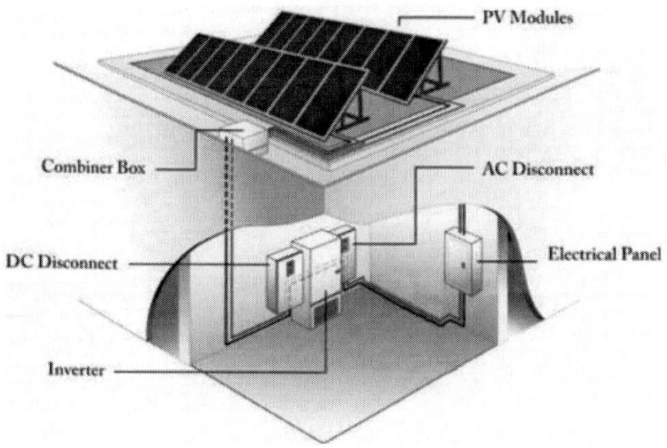

Figure 16. PV system components.

A monitoring system is a common component in PV installations. Monitoring allows tracking of solar system performance over the life of the system and makes it easy to identify systems that are performing below expectations and may be in need of maintenance. Monitoring systems require an Internet connection in the room that contains the inverter. Internet access to this space should be considered in the building design phase.

Consider Special Loads (PV)

If there are auxiliary systems that require uninterrupted power (i.e., security systems, fire alarms) consider available storage options. In the event of a power outage, the storage system would feed power to the system for a pre-determined amount of time. Figure 17 shows a possible configuration of components that will continue to deliver power to the critical load in the case of a grid power outage. Emergency call centers, fire and police departments, schools, and emergency shelters are often good candidates for battery-supported PV systems.

Sizing the System (PV)

The size of the system will depend on the solar resource of the proposed location. The solar resource of the site should be researched to determine whether the site has sufficient resource to facilitate a PV system. Figure 18 shows a solar resource map of the United States. Site-specific hourly solar radiation can be found using the National Solar Radiation Database.

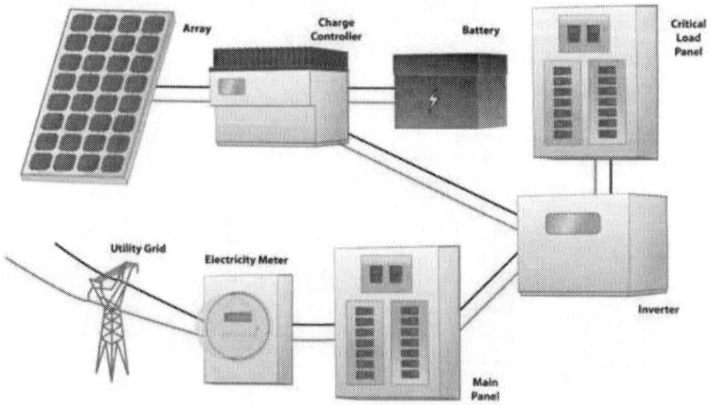

Figure 17. PV system with battery storage and grid connection.

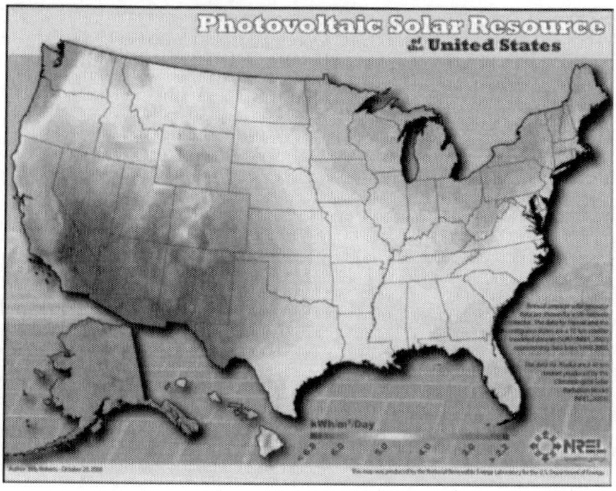

Figure 18. Solar resource map of the United States.

Yearly Data: http://rredc.nrel.gov/solar/old_data/nsrdb/1991-2005/hourly/list_by_state.html.

Typical Meteorological Year (TMY) Data: http://rredc.nrel.gov/solar/old_data/nsrdb/1991-2005/tmy3/by_state_and_city.html.

When a building is being prepared for a solar installation, it is important to know the size of the proposed system so that all the correct preparations can be made. Typically, PV systems are designed only to provide a portion of the energy usage required by a site. In most places, it is not desirable to produce more energy than can be used on the site annually due to the low utility rate that is paid for electricity. For this reason systems are typically designed to meet 80% of the site energy usage to ensure that the system does not produce more energy than the site can consume, even during low energy consumption years. The National Renewable Energy Laboratory developed a Web-based calculation tool called PVWatts that can be used to size a PV system by finding the annual energy production of the system. (http://rredc.nrel.gov/solar/codes_algs/PVWATTS/version1/). PVWatts uses the system size, orientation, location, and collector efficiency information to calculate the system output. If the annual energy production is not the desired amount for the specific site, the system size can be adjusted.

In order to determine what size PV system will fit in a designated area, the system type must be known. Each PV system will have a different area requirement, but typically similar technologies have similar footprint requirements per unit size. The following numbers can be used to find the approximate system size that can be placed in a given space.

- Crystalline PV System: 8-11watts/square foot
- Thin Film PV System: 5 watts/square foot
- HE Crystalline System: 17-18 watts/square foot

Conclusion

A few simple considerations when designing buildings will facilitate a smooth and cost effective transition to solar later in the building's life. In cases in which solar is not economically feasible during the initial construction phase, making the structure solar ready will help reduce the carbon footprint of the building over its lifetime and lower power costs when the solar system is installed. Furthermore, a solar ready building will position the building owner to take advantage of falling renewable energy prices in the future. Implementing policy that requires some or all new construction to be solar ready is a simple way community leaders can promote solar in their jurisdictions. With the aid of the Solar Ready Building Planning Guide, city planners, policymakers, and developers will be able to lay the foundation for Solar Cities.

End Note

[1] www.solar-rating.org

INDEX